George Hazen French

The Butterflies of the Eastern United States

George Hazen French

The Butterflies of the Eastern United States

ISBN/EAN: 9783743383807

Manufactured in Europe, USA, Canada, Australia, Japa

Cover: Foto ©berggeist007 / pixelio.de

Manufactured and distributed by brebook publishing software (www.brebook.com)

George Hazen French

The Butterflies of the Eastern United States

THE

BUTTERFLIES

OF THE

EASTERN UNITED STATES.

FOR THE USE OF CLASSES IN ZOOLOGY AND PRIVATE STUDENTS.

BY

G. H. FRENCH, A.M.,

PROFESSOR OF NATURAL HISTORY AND CURATOR IN THE SOUTHERN ILLINOIS
NORMAL UNIVERSITY.

\

PHILADELPHIA:
J. B. LIPPINCOTT COMPANY.
1886.

PREFACE.

For a number of years the writer of these pages has recognized the necessity of some form of manual to be placed in the hands of students in zoology, to enable them to identify the animals that should come before them for examination. Jordan's "Manual of Vertebrates" supplies this demand as to the vertebrate animals, but there are many other specimens of interest within the reach of every student that do not belong to this great branch of the animal kingdom, among the most attractive of which a. ; butterflies. Several years ago, analytical tables of the butterflies of Illinois were prepared and published for the use of our classes in zoology. These tables, followed by others on the moths, notwithstanding many imperfections, served so good a purpose in the class-room, and were sought by so many private students, that the preparation of a more extended work on the Butterflies of the Eastern United States has been undertaken. The work embraces a brief description of the several stages of butterflies, methods of capture and preservation, an analytical key, and a more complete description of all the species that have been found in this region. In the last part the preparatory stages are given so far as they are known. These preparatory stages are often essential to a proper understanding of the relation that species bear to one another, besides adding much to the interest of the study of butterflies.

The locality represented in this work is shown on the map on the opposite page, being all east of the western boundaries of Minnesota, Iowa, Missouri, Arkansas, and Louisiana, as indicated by the heavy line. This differs a little from what is denominated the Eastern or Atlantic Province on the zoo-geographical map by Dr. A. S. Packard, Jr., in the third report of the United States Entomological Commission, in that the western boundary is by State lines, instead of following a more sinuous line caused by variations of elevation, etc., and the whole of Florida and the New England States are included, but the portion of Canada included in the map referred to is omitted here. For several reasons it was found more convenient to take the boundaries as here given, making the field represented essentially the same.

I would gratefully acknowledge here the valuable aid I have received from Mr. William II. Edwards, of Coalburgh, West Virginia, in the loan of specimens for description, in the free use of his writings, from which nearly all the descriptions of the preparatory stages have been taken, in the use of advance-sheets of his new catalogue of the "Butterflies of North America," for the purpose of getting localities and arrangement of species, and in many suggestions kindly given. Mr. C. E. Worthington, of Chicago, also loaned me specimens for description, thus aiding me much. I wish to acknowledge also the aid and encouragement I have

received from Dr. Robert Allyn, president of the
Southern Illinois Normal University, not only in the

preparation of this volume, but in the interest of science
generally. Dr. S. H. Peabody, regent of the Illinois
Industrial University, also has thanks for revising a list

of the accentuated names of the butterflies of the Eastern United States. Last, but not least, I would acknowledge valuable aid and encouragement from my wife, who has been the companion of my studies in natural history for many years, making it possible for me, at this time, to prepare this volume. In the few instances where I have not had specimens for description, the back volumes of the "Canadian Entomologist" and the American Entomological Society's publications, Professor Fernald's "Butterflies of Maine," and some other works, have been used.

Electrotypes for illustrating this volume have been received from the following persons:

From Professor C. V. Riley, Nos. 8, 9, 10, 11, 12, 13, 23, 24, 25, 26, 27, 28, 29, 30, 36, 37, 38, 39, 59, 60, 61, 62, 63, 64, 65, 66, 67, 68, 69, 91, 92, 93.

From Henry Holt & Co., of New York, the figures from Scudder's Butterflies, Nos. 1, 2, 3, 5, 7, 33, 41, 42, 45, 51, 52, 54, 55, 71, 73, 75, 76, 77, 78, 79, 80, 81, 82, 83, 87, 88, 89, 90.

From Dr. A. S. Packard, Jr., the figures from "Guide to the Study of Insects," Nos. 4, 6, 40, 44, 46, 47, 48, 49, 53, 72, 84.

From J. B. Lippincott Company, Philadelphia, the figures from Saunders's "Insects Injurious to Fruits," Nos. 17, 18, 19, 20, 21, 22, 58, 74.

From A. A. Tenney, the figures from Tenney's "Elements of Zoology," Nos. 14, 15, 16.

All the rest were made specially for this work by the St. Louis Type Foundry Company, from photographs taken by Mr. A. Hudson, of this place.

In the arrangement of species and nomenclature,

Edwards's "New Catalogue of the Butterflies of North America" has been followed.

It has been suggested that I should give, in addition to the table for tracing species to their description, a table of larvæ. In my opinion our present knowledge is not sufficient to make a satisfactory one. Though much is known of the preparatory stages of our butterflies, there are still too many gaps : these, however, are being filled up, so that in a few years a table can be given that will be more satisfactory than any that could be prepared now.

G. H. FRENCH.

CARBONDALE, ILL., June 8, 1885.

CONTENTS.

10 CONTENTS.

LIST OF ILLUSTRATIONS.

2

THE BUTTERFLIES

EASTERN UNITED STATES.

CLASSIFICATION.

MOST naturalists of this country divide insects into seven orders, or, according to some, suborders, as follows :

HYMENOPTERA, or membrane-winged insects, including bees, wasps, ants, ichneumon-flies, saw-flies, etc.

LEPIDOPTERA, or scaly-winged insects, including butterflies and moths.

DIPTERA, or two-winged insects, including the true flies of all kinds.

COLEOPTERA, or sheath-winged insects, including all beetles.

HEMIPTERA, or half-membrane-winged insects, including the true bugs, cicadas, etc.

ORTHOPTERA, or straight-winged insects, including grasshoppers, katydids, crickets, etc.

NEUROPTERA, or nerve-winged insects, including dragon-flies, ant-lions, etc.

The Lepidoptera have four membranous wings covered

15

with scales, that overlap one another and easily rub off.
They are divided into two somewhat natural divisions,—
butterflies and moths,—which may be known from each
other by the antennæ, but by the practical entomologist
are readily distinguished by other characters. The antennæ, two slender organs projecting in front from the
upper part of the head, are filiform, and terminate in a
knob in butterflies, while in the moths, whatever their
shape, they do not terminate in a knob. Butterflies fly
in the daytime, while most of the moths fly at night
or just at the close of day. The first five families of
this order are known as butterflies, and all the others as
moths.

Both moths and butterflies have four distinct periods
of existence,—the egg, the larva, the chrysalis, or pupa,
and the imago, or perfect insect.

THE EGG.

The eggs are more or less globular, cone-shaped or
spindle-shaped. To the unaided eye they usually appear
to be smooth, but under the lens they present various
markings that are constant for a given species, but vary
with different species or in different groups. Those
belonging to the different genera of the subfamily
Pierinæ are all of one general shape, as has been shown
by Mr. W. H. Edwards. They are long, slender, sub-
conic or spindle-shaped, and set on end, but differently
marked in different genera. This may be seen by referring to the egg of *Pieris Oleracea*, Fig. 27. The
eggs of Danais, Heliconia, and Agraulis have each their
own pattern. All Argynnis eggs, whether of the large
or the small species, are thimble-shaped. "So Melitæa,

Phyciodes, Limenitis, Apatura, Paphia, Satyrus, Neo-nympha, and Chionobas may all be distinguished as readily by the egg as by the butterflies." And the same may be said of all the other genera so far as their eggs are known. It has been shown that both the larvæ and the eggs of the different species of a genus bear the same relation to each other that the imagines, or per-fect insects, do, and this relation or congruence renders the study of the preparatory stages important, if we would know the true relation that different species bear to one another, or in fact the position of different genera. "Most eggs," Mr. Edwards says, "are green when laid, yellowish, as in Pieris, Colias, and many Papilios, bluish, as in Grapta, grayish, as in Limenitis. Lycæna has a deep-green surface concealed by a white net-work, but which can be peeled off. Parnassius is white; *Pholisora Catullus* is brown; the Hesperian eggs, as a rule, are white. Many eggs turn red a few hours after deposition, as Colias, Anthocharis, and *Melitæa Phaton.* And all these, as well as most other species, change to black before hatching, as the dark larva can be seen through the transparent shell."

Many eggs are ribbed longitudinally, with transverse striæ between the ribs. In some these ribs run in irregular lines, making an irregular net-work of cells; in others they are regular, the net-work presenting the appearance of a series of parallelograms, as in Fig. 27.

In some cases the eggs are not ribbed, but are covered with a series of irregular pits, as in Fig. 75. In some the ribs run from base to apex, varying or not in promi-nence; in others they gradually diminish before reaching the base, leaving that part perfectly smooth. The egg of

b   2*

Limenitis Disippus presents on the ridges of the cells numerous little filaments, giving the egg a bristly appearance (see Fig. 60). In all these there is a cluster of irregular cells at the summit or apex that differ from the others, often being situated at the bottom of a cavity or depression. This portion of the egg is known as the micropyle.

Fia. 1.

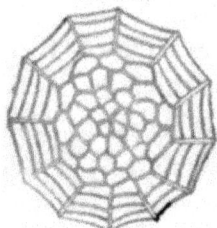

Micropyle of egg of Colias Philodice. × 60.

Butterflies vary in their manner of depositing their eggs. Some place them singly on the leaves of their food-plant, while others lay them in clusters, from a dozen to a hundred in a cluster. In the case of *Vanessa Antiopa*, Fig. 54, they are placed around a small twig of willow. The Papilios, with the exception of *P. Philenor*, deposit their eggs singly. The Graptas lay their eggs in strings or singly. Usually the manner of depositing the eggs may be taken as an index of the larva's being gregarious or feeding singly.

The time of the egg period, or the time from deposition to hatching, varies in different species, depending somewhat upon the temperature. In some species they hatch in from three to four days, as in Grapta, Colias, and Pieris. The Papilios and *Danais Archippus* hatch in from four to six days; *Phyciodes Tharos*, in from four to seven days; *P. Nycteis*, in twelve; *Limenitis Disippus*, in from five to eight; *Argynnis Diana*, in fifteen; *A. Cybele, Aphrodite, Alcestis*, and *Atlantis*, in from fifteen to twenty, etc.; while, according to Mr. S. H. Scudder, there are some species that pass the winter in the egg state.

THE LARVA.

The larva, or caterpillar, is elongated or somewhat worm-like, usually plainly separable into thirteen joints or segments, the first of which is the head. Joints two, three, and four have each a pair of short legs, the rudiments of the legs of the perfect butterfly. Joints seven, eight, nine, ten, and thirteen have each a pair of membranous legs, the end of each armed with a circle of minute hooks, as seen in Fig. 2. By means of these hooks the larva is enabled to grasp firmly any object which is not too smooth, as the surface of glass. In this case the larva first spins a covering of silk over the glass, and then walks over it easily. These legs are called prop-legs, or, as it is more often abbreviated, prolegs. These disappear at the close of the larval period, when the larva changes to a chrysalis.

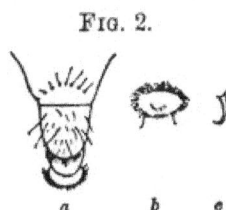

FIG. 2.

From larva of V. Antiopa. *a*, proleg, × 4½; *b*, circlet of hooks at end of proleg, × 6; *c*, one of the hooks, × 12.

On each side of the body are nine oval stigmata, or breathing-pores, often called spiracles. These are situated in joints two and five to twelve inclusive. These stigmata open into a series of air-tubes which ramify through the system, each stigma leading to a single trunk of the system. Close to the origin of this trunk a large air-canal runs along each side of the body, connecting all the trunks of one side. Joints three and four, having no stigmata, receive their branches of the system of air-passages from this trunk. Like the air-passages in the lungs of the vertebrate animals, these tracheæ continue to

divide and subdivide till the minute tubes penetrate all
parts of the body, especially all parts of the circulatory
system. It is in these ultimate divisions that the inter-
change of gases takes place which constitutes the purifi-
cation of the blood, or circulating fluid.

The head is of a rounded or oval form, and has a harder
covering than the other parts of the body. When the
larva is first hatched the head is nearly globular, divided
down the front by a suture which forks about the middle.
As the larva approaches maturity the head usually
changes in shape, assuming the characters that are
peculiar to the species. The lower edge of the little
triangular piece which stands between the forks of the
frontal suture supports a little membrane, the labrum,

FIG. 3.

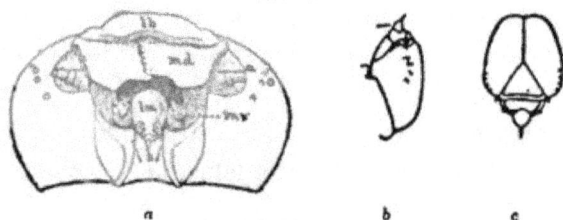

a, Head of larva of Danais Archippus from beneath, ✕ 10 : lb, labrum; md, mandi-
ble; mx, maxilla, with two palpi; lm, labium, with one pair of palpi; s, spinneret;
a, antennæ; o, ocelli; b, side view, and c, front view, ✕ 3. (Scudder, after Burgess.)

or upper lip, and back of this are two stout biting jaws,
or mandibles, with serrated edges, that work laterally (see
Fig. 3). The mouth lies between these jaws, and back of
them are the secondary jaws, or maxillæ, which in many
insects have a movement similar to that of the mandibles,
but they do not in the butterfly larvæ. They consist
of a pair of fleshy prominences, and each of them has

two feelers, called palpi, the outer composed of several joints, the inner of only one. Between and partly below or back of the maxillæ is the labium, or under lip, being more like the maxillæ than like the upper lip. This bears on each side at the tip a small jointed appendage : these appendages are termed the labial palpi. Between these is another appendage, tubular, which is similar to the spinneret of the spider, and from which the caterpillar spins a web over smooth surfaces as a support for its feet in walking, and the silk it fastens its feet to in moulting and in changing to a chrysalis. Back of the jaws, somewhat in the form of a crescent, are the eyes, or ocelli, five or six on each side.

The bodies of different caterpillars differ greatly in their external covering as well as in shape. Some appear to be naked, but even these are covered with a delicate pile; others have simple or compound spines or tubercles, usually arranged in longitudinal rows with a definite number to each joint, generally beginning with joint three, or the second thoracic segment.

Usually the larvæ of butterflies are cylindrical, in some a little enlarged at or near the middle; in others, as in some Papilios, the thoracic segments are enlarged, and at times assume shapes peculiarly their own. In such cases the head is often smaller than the succeeding joints, and when at rest is drawn back, as it were, into the joint behind. In others the second segment is smaller than the head, as with many of the Hesperians.

Most butterfly larvæ have the thoracic and abdominal legs as given at first, but in some, as the Lycænidæ, the prolegs are very small, and the caterpillar seems to glide over surfaces instead of walking, the under side being a muscular pad, by whose **expansions** and contractions the larva moves.

The Papilios have in the upper **part** of the second segment a peculiar V-shaped extensile appendage, known as a "scent-organ," **or** osmateria, which they protrude from a transverse slit when disturbed, but which is at other times concealed. This organ is without doubt used as a defence, the disagreeable odor emitted repelling enemies. In some of the Lycænidæ the posterior part of the body has **extensile organs that secrete a sweet fluid which is eaten greedily by ants. These in turn guard the larvæ against the attacks of ichneumon-flies, very much as they protect plant-lice from the attacks of enemies for the sweet fluid they get from the honey-tubes of the lice.**

In passing from the eggs to the full-grown larvæ, caterpillars **moult or shed** their **skins** from **four to** five times. At each **moult they not only** come out in a skin that is larger than **the old one, which thus** permits further **growth,** but **the color and other markings** are usually **changed.**

In habits of **feeding each species has its** larval peculiarities. Some **feed singly, as the** larvæ of *Grapta Comma,* on the **under side of a** hop- or **nettle-leaf. Some stitch together the edges of a leaf,** making a more **or less closed retreat;** others feed **on** the surface without **any attempt at** concealment, as *Papilio Cresphontes,* but here **the color** and shape so mimic an object which would **be distasteful** to birds that **it is not molested** by them. The young larvæ of **Apatura are gregarious, but are** not protected **by a web. After the third** moult they scatter, and the rest **of the time** are solitary. *Melitœa Phaeton* larvæ **make a web,** within which they feed **till after the third** moult, when they close every place of egress and pass the winter in the web. In the spring they leave the web and bask in the sunshine on the leaves.

THE CHRYSALIS, OR PUPA.

When ready to change to a chrysalis, the larva seeks some retreat, if it be one of the species that does not pupate in a web or a cluster of leaves, where it prepares to change to the pupal or quiescent period. In the Papilionidæ and Lycænidæ this preparation consists in spinning a button of silk on the under side, or side, of some object, within which it entangles the hooks of the anal feet. Then a loop of silk is woven from side to side that will support the body a little in front of the middle, in which the body is allowed to rest, held in place by the anal feet. Soon the skin is shed, and the chrysalis appears limp and pale, but as the moisture is evaporated the outside hardens, and it assumes a shape and color peculiar to the species. In others, as the Nymphalidæ, the button of silk is spun and the anal feet are entangled in it, but the front part of the body hangs down without the loop of silk to support it. The anterior part bends like a fish-hook, after which the skin is shed and the chrysalis suspended by the anal hooks.

In the larva there was but little distinction of parts, as head, thorax, and abdomen. In the chrysalis there is more of a division of these parts, the head and thorax being united, but the abdomen readily separable.

In moths the head part of the chrysalis is usually rounded, but in butterflies, especially some of the large Papilios, the cephalo-thorax bears several prominences and ridges, often continued along the abdomen. The ventral part contains cases for the wings, antennæ, tongue, palpi, and legs. The wing-cases extend back several joints on the abdomen, often as far as the posterior edge of the fifth joint. Between the wing-cases, and extending back varying distances, are four other cases, the centre the tongue-case, next cases for the anterior pair of legs, next cases for the middle pair, and outside of these the antennæ-cases. The base of the tongue-case is larger than that organ is, but the expansion of this part is used as a covering for the palpi and to fill up the space between the legs. The posterior pair of legs are folded beneath the wings, and are not shown in the chrysalis by any case. When the chrysalis has become dry and hard, these cases are inseparable; but when the larva skin is first cast off and the parts are soft, they may be separated by a sharp-pointed instrument.

The anterior part of the head may be rounded, but more often it ends in two conical points or a single point. Just back of this part, and near the base of the antennæ, is a smooth, crescent-shaped belt which corresponds in position to the ocelli of the larva. The use of this is not fully known, though it is without doubt a covering for the eyes. Back of these parts, on what is called the pronotum, is often another elevation with ridges running along the sides. In Limenitis there is a prominent, rounded elevation back of the mesonotum. In some species the elevations and depressions are too complicated for general description.

The abdomen is more or less conical, tapering towards the anal joint, which ends in a complicated series of hooks known as the cremaster. These hooks are fitted for

being fastened into the button of silk to which the anal
legs were attached before changing from the larva to the
pupa state. The joints back of the wing-cases are more
or less movable. In the chrysalides of the Papilios the
lateral ridges of the cephalo-thorax are continuous to the
cremaster; in others, as some of the Nymphalidæ, there
are rows of tubercles or spines. In all cases the abdomen
contains a row of stigmata on either side which correspond to those in the larva, except the anterior, whose
places are covered by the wing-cases.

The outside covering of the chrysalis is a fine, horny
substance known as chitine, the same as forms the hard
parts of all insects. In most of the pupæ this is in color
greenish, yellowish gray, or some shade of brown, sometimes ornamented with bright metallic spots.

While the pupal period seems externally to be one of
inactivity, internally great changes are going on,—the
preparation for a change from the worm-like caterpillar,
which can only creep or slide over a leaf or twig, to the
airy and graceful butterfly. The time in which this
change takes place, the pupal period, varies greatly,
ranging from six or seven days to several months, as
with those that hibernate in this state; but about fourteen days is the usual time.

When the pupal period draws to a close, the pupa-case
is burst open on the dorsal part of the cephalo-thorax,
and the butterfly, or imago, emerges with all its parts
limp and moist. This bursting of the case is accom-
plished partly by the moisture that is exuded from the
interior for the purpose of softening the inner integu-
ment of the shell, and partly, it seems, by favorable
atmospheric conditions, as the moist atmosphere of a

damp evening or a warm rain is more conducive to this change than dry weather.

THE IMAGO.

After emerging from the pupa-shell the butterfly finds some place, often the pupa-case, where it may rest with the body hanging downward, and after a moment's delay, as if for rest, it proceeds to expand the wings, which were before not larger than the finger-nail. This is done partly by their own weight, but mostly by forcing air into the hollow veins that constitute the framework of the wings. After the wings are expanded to their full size, the insect remains till they are fully dry before it flies away.

A butterfly has three principal divisions of the body,—the head, thorax, and abdomen. The head is more or less globular, and contains the mouth parts, the eyes, and the antennæ as its principal divisions. On either side of the head are the eyes, two convex hemispheres that are made up of a great number of small eyes or facets, the whole on each side of the head being known as a compound eye. In some species as many as three thousand six hundred and fifty facets have been counted in a single eye. Each of these is hexagonal in shape (see Fig. 4), and contains all the parts of a perfect eye. The surface of this compound eye may be smooth, or moderately covered with short hairs, which are situated between the facets. Some of the moths have besides the compound eyes ocelli above these, and it is said one species of the butterflies has one on each side; but aside from this these insects have only the compound eyes.

Fig. 4.

Facets of a compound eye.

Above the eyes are the antennæ, two long, jointed organs, each composed of many joints, which may be divided into three groups,—those of the base, the stalk, and the club. The two joints composing the base are larger than the others; the stalk is merely a jointed thread; the club has the joints shorter and broader. In some cases the antennæ are bare, in others they are more or less clothed with scales. The use of these organs is not fully known, but they are supposed by many to be organs of hearing. In the upper part of the club are microscopic pits connecting with nerves, showing that the antennæ are sense-organs; and it is probable they are not connected with the same sense in all insects. In some beetles, and some grasshoppers, ants, and bees, the sense is without much doubt one of touch; in some moths it seems to be one of smell.

On the under side of the head are the mouth parts. These consist first of a three-jointed pair of palpi, which are densely covered with hair-like scales, and which project outward and often curve upward more or less closely to the front of the head. Between the palpi, and attached to the head near the base of them, is the proboscis, or tongue (see Fig. 5). This is a long, tapering, horny tube, through which the insect sucks or draws up fluid substances from flowers or other objects. When at rest, the tongue is coiled backward between the palpi like a watch-spring; when uncoiled, it is often as long as the body of the insect. It consists of two lateral halves united down

Fig. 5.

Head of *E. Tityrus*, showing tongue and one antenna.

the middle, each of which is composed of a great number of rings, convex on the outer part and concave on the inner, and the tube is formed by the union of these concave surfaces. The head is clothed with a dense coat of hair-like scales, often spoken of as hairs, and the arrangement of some of these is of value in determining genera or species.

The thorax is connected or joined to the head by the neck, and bears the legs and wings. It consists of three joints, to each of which is attached on the under side a pair of legs, but only the two posterior joints are furnished each with a pair of wings. Each leg is composed of a basal joint, called the coxa, at the end of which is a small piece called the trochanter. Beyond this is the femur, the longest joint of the leg; attached to this is the tibia, followed by the tarsi, or foot, which consists of five joints placed end to end, the last of which usually has a pair of curved claws. The middle and hind tibiæ usually have a pair of spurs at the end of each, and are sometimes more or less armed with spines. The hind tibiæ in some species have an additional pair of spurs near the middle. In some species the front tibiæ have an appendage on the middle of each, called an epiphysis.

In one family, the Nymphalidæ, the front legs are so much aborted as to be of no service in walking; and such are said to be four-footed butterflies. In the other families the pair of fore legs is directed forward, and the middle and hind legs backward; but in this family the second pair of legs is directed forward.

The first ring of the thorax, the prothoracic, is smaller than the others, and its only appendages are a series of scales arising from the upper side, forming the collar, and on each side a small, scaly piece covering the base of the fore wings, and known as the shoulder-tuft, lappet, or pterygoid. The second and third thoracic joints bear each a pair of wings. These are composed of membranes supported by a framework of slender, tapering tubes between the membranes. The fore wings are the largest, triangular in general outline, while the hind wings are more or less rounded or square. The veins or framework are nominally five principal veins,—the costal, subcostal, median, submedian, and internal. The first two are close together near the front edge of the wing, and form the costa (see Fig. 6). The median passes through the middle from the base to near the outer third, where it usually joins the sub-

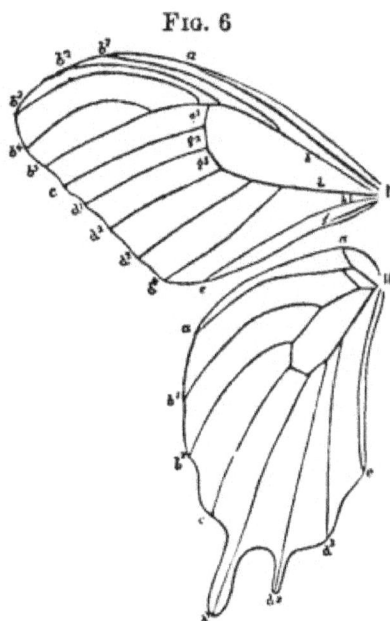

Fig. 6

Fore and hind wing of a butterfly: 1, fore wing; *a*, costal vein; *b*, subcostal vein; *b* 1, *b* 2, *b* 3, *b* 4, *b* 5, five subcostal veinlets; *c*, independent vein; *d*, median vein; *d* 1, *d* 2, *d* 3, *d* 4, four median veinlets; *e*, submedian vein; *f*, internal vein; *b* and *d* are situated in the discal cell; *g* 1, *g* 2, *g* 3, the upper, middle, and lower discal veinlets; 1 1, hind wing (the lettering the same).

costal by a cross-vein; and from this and the cross-vein are given off several branches, the subcostal also being branched on its upper side, more in the fore wings than in the hind. The area between the subcostal and median veins is known as the discal cell, or the cell, the branches of the median vein as the median venules or veinlets, the branches of the subcostal as the subcostal venules or veinlets, the branches from the cross-vein as the discal venules or veinlets.   The space through which these venules pass is sometimes spoken of as the discal space, or disk.   The submedian and internal veins occupy the area below the median, the latter being short and sometimes wanting.

The arrangement of these veins—called by some authors nerves and nervures—is of value in classification, and they also serve to locate markings which rest either near or upon them.   When the wings are expanded (and that is presumed to be the case in the following descriptions of species and in the key), the front edge is called the costa, the part next to the body the base, the edge farthest from the body the outer or terminal margin, the part opposite the costa the posterior or hind margin (the inner margin of some authors).   The angle between the costa and the outer margin is called the apex ; the one between the outer and hind margin may be known as the posterior angle when applied to the fore wings. The hind wings have the costa, outer margin, and apex the same as the fore wings, the latter being sometimes spoken of as the outer angle, but the part of the hind wing next to the body is called the internal or inner margin, and the angle at the end of this the anal angle.

In Europe, and to some extent in this country, a system of numbering the veins has been adopted.   The plan is to number them in order at their termination along the margin of the wing, without regard to their length.   By this plan the one extending from the base of the wing below the median would be called 1, the first or lower branch of the median 2, the second branch 3, and so on round the outer margin and costa to the costal vein, which will have the highest number. If, however, there are two veins below the median, the submedian and internal, the first is called 1 a, the second 1 b. The same system is observed with the hind wings.

The membranes of the wings are concealed beneath a covering of minute colored scales. The membrane itself is not colored, the colors of the wing being due to the various hues of the scales. These are arranged in regular rows (see Fig. 7), and lap over one another like shingles on a roof. The scales are modified hairs, and are of various shapes. The basal end by which the scale is attached to the wing comes more or

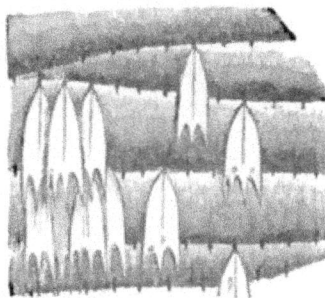

FIG. 7.

Section of butterfly-wing showing how the scales are attached.

less abruptly to a point; but the other end varies, being rounded or variously toothed or pointed. This covering has gained for these insects the scientific name Lepidoptera, from two Greek words which signify scaly-wings.

These scales cover both the upper and the lower surface, and they are usually of a different color below from what they are above. Sometimes this difference is merely a difference in shade of the same general color, at other times it is more than that, as in the Papilios, etc. The two sexes are often different on the upper surface, but are more nearly alike beneath, as in many of the Pamphilas.

The abdomen is either oval, as in Papilio, Vanessa,

etc., or more slender, as in Pieris, or nearly conical, as
in some of the Hesperidæ. It consists of eight or nine
segments, each furnished with a spiracle on each side.
The digestive system, which in the larva state was an
alimentary canal, consisting of a cylindrical muscular
tube extending from one end of the body to the other,
enlarging in some places and contracting in others, so
as to be naturally divided into œsophagus, stomach, and
intestines, now has changed into a more slender, tortuous
tube twice the length of the body. The respiratory
system is similar to that in the larva state. The nervous system consists of seven ganglia, while in the larva
there were eleven. The reduction is due to the fusion
during the pupa state of those in the anterior part of
the body, forming two thoracic ganglia, which distribute
nerves to the legs and the muscles of the wings. The
ganglia in the head and abdomen give off fibres to the
various organs of these parts, each ganglion serving as
a brain to the part in which it is located, but at the
same time communicating with the other ganglia by
nervous filaments.

HABITS OF BUTTERFLIES

Butterflies are day-flyers. They rejoice in the warm
sunshine, few being seen on the wing if the weather be
cloudy with a cold wind. On the side of a mountain as
the sun was setting, throwing different portions into the
shadow from the base to the top, the writer has seen the
butterflies fly from cluster to cluster of flowers up the
acclivity, going just fast enough to keep in the sunshine. The kinds that are to be found only in the
woods will be seen flitting about in a patch of sunshine

where the sun shines through a break in the trees, sipping sweets from the flowers or basking on a leaf; but if some other patch of sunshine is sought, it is by nearly direct flight. It is true such butterflies as *Debis Portlandia* are almost habitually in the shade; but even they are more active on sunshiny days than when the sky is overcast with clouds.

The direction of the wind seems to affect all insect life. Though the sun may shine in a cloudless sky, if the wind blow moderately strong from the northwest, butterflies take to the wing but little; and there is more in this than the fact that a wind prevents their flying with ease. A much stronger south wind would tempt them forth and cause them to be blown about where the wind was strongest, but behind some hill or sheltering wood they would be found more at their ease.

The habits of different species in the places they frequent vary greatly. There are a few species, as *Colias Philodice*, *Danais Archippus*, and a few others, that are to be found everywhere within the limits of their range, in wood and field, town and country. *Papilio Asterias* is another species that has a wide range, while *P. Troilus*, *Ajax*, *Philenor*, and *Cresphontes* are confined more to the open woods, where they may be seen in search of their food-plants, or hovering over the flowers of some Vernonia or Eupatorium, or slaking their thirst at a damp place in the road. With wind and weather favorable, these may often be seen on flowers at a distance from the woods. *Callidryas Eubule* and *Sennæ*, when they occur in this region, are to be found in the fields or open woods; but they fly rapidly, stopping for a moment on flowers, seemingly as though migrating. From this

c

manner of flying, which is usually in a north or south general direction, the writer has thought that they did not breed here, but that the larvæ were to be found farther south; and this has in a measure been confirmed by never finding any larvæ or eggs on their food-plants.

The food-plant of a species determines to some extent the places of its resort. Cabbage and turnips being largely the food-plants of *Pieris Rapæ*, this species will be found more about gardens and fields where these plants are grown. The tame and prairie grasses furnishing food for the different forms of *Satyrus Alope*, this species will be found in meadows and prairies; while the Neonymphas and *Debis Portlandia*, feeding more on the grasses growing in shady woods, may be sought in these places. The Neonymphas fly low and with a jerking motion, unless disturbed, but Debis has a different flight. The male selects some tree, on whose trunk he may be found, darting out upon every intruder, large or small, to return again to his post; the female being near by, perched upon a blade of grass or a leaf. The Theclas are to be found in some open wood or on bushes along the border of a clearing. They rest upon the sunny side of a bush on a leaf, frequently flitting out and back again to the same or an adjoining leaf; and *Feniseca Tarquinius* has a similar habit. The Lycænas are to be found more about grasses and flowers, or hovering over some moist place in the path or about some pool or small stream. The different species of Pyrameis or Junonia are often to be found in a path or road, from which they will fly up to alight a short distance ahead, flying past you after this is repeated a few times. *Limenitis Disippus* has a similar habit, being found not far from some clump of willows. *L. Ursula* is more often found in or near the woods on the lower leaves of some tree or shrub, or sipping moisture from a mud-hole. The Pamphilas are essentially grass insects, and are seen more frequently about the rank growths of semi-water-grasses of a swamp than in any other place, except a blossoming clover-field. In spring the wild plums and judas-trees form a resort for several species of Eudamus, Nisoniades, and Papilios, as well as for many other insects.

Some species flock together in great numbers, especially after they have multiplied to a great extent, as *Danais Archippus;* others are seldom to be found except in pairs, as *Debis Portlandia* already spoken of, and *Paphia Troglodyta.* The latter, instead of sitting upon the trunk of a tree, takes position on a leaf, a stick, or a stump, where he stands guard over his mate, chasing away every intruder and returning again to the same place. When the sun sinks in the west, or the sky becomes overcast with clouds, the butterflies prepare for the night's sleep. In doing this, they usually attach themselves to the under side of a leaf, with the wings folded back to back, and the fore wings thrown back so as to be partly covered by the hind wings. Many species are of such colors on the under side that in this position they are not conspicuous, the colors simulating those of the surrounding objects. The coppery brown of the under side of Paphia closely resembles that of a dead oak-leaf, and so do the dull browns of Satyrus, Neonympha, and others, though some are more variegated. The writer has frequently seen *Argynnis Cybele* fly about several low trees and try several leaves before finding one to its liking. Butterflies will sometimes do this to

avoid danger.   A female *Calidryas Sennœ* was struck at
by the writer with a net as it was passing on the wing.
It dodged the net, but at once turned from its course and
flew to a small oak-bush, where it settled, in an attitude
of repose, on the under side of a leaf, from which it was
taken by the hand.

Some species hibernate in the butterfly, or imago, state,
as *Vanessa Antiopa* and some of the Graptas. If a chip
is cut from a tree in the forest so that the sap flows a
little, these butterflies may be seen late in the fall, when-
ever the weather is mild, sipping at such a place.   As it
becomes cooler they retire to some sheltered place, where
they anchor themselves by the hooks in their feet and
become lethargic, remaining there till the warmth of
spring arouses them from their slumbers. The wounded
trees, fresh-cut stumps, and early flowers furnish them
the food their system demands in the spring; and in
due time the eggs are deposited for the new generation.

COLLECTING BUTTERFLIES.

This may be considered under two heads,—collecting
the adult imagines
and rearing them
from the eggs or
larvæ.   In the first
a few implements
are essential, though
they need not be ex-
pensive,—a net and
a poison-bottle.   A
net to be used easily should be made as light as possible,
though it must be stout enough to be serviceable.   It

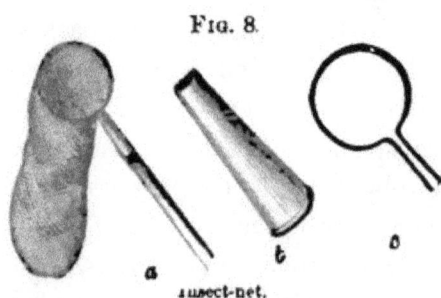

Fig. 8.

insect-net.

may consist of three parts, the bag, hoop, and handle, as shown in Fig. 8. The hoop, *c*, should be made of about No. 8 wire, and nine inches in diameter will be a convenient size. Any tinsmith can make this. The wire, after being bent in the form of a circle, should have the two ends bent out so that they may come together as in *c*, though not left so long, an inch and a half being long enough. The second part of the hoop, shown at *b*, consists of a tin ferrule which may be a socket for the reception of the handle. This should be four and a half inches long, three-fourths of an inch wide at the large end, and tapering down so that the two ends of the wire when placed close together will just fit in. Place these in the small end of the ferrule till the tin comes against the circle, and fasten with solder. Many use a patent brass socket with an adjustable wire fastened with a screw, but I find these heavier than the one here described, besides being more expensive, this not costing more than from fifteen to twenty-five cents. The bag part of the net, *a*, should be made of some strong but light material, such as " Swiss," though mosquito bar will do very good service. If the material used is a yard in width, this may be taken for the length of the bag, and the dimensions the other way so much as will go round the hoop. The bottom of the bag should be rounded, the cloth of the other end put over the wire, and over this a piece of strong muslin, and the whole sewed close to the wire.

For a handle a stick about the size of a walking-cane will answer, or one two and a half feet in length, made of some light but stout wood. Black walnut and ash are preferable to any of the softer woods, as they are not so easily broken. The handle should not be more than

4

three-fourths of an inch in diameter, and should taper a
little at one end so as to fit into the socket of the hoop.
A net made in this way is light enough to be used easily
by even a child without straining the wrist, and yet is
strong enough for all ordinary purposes. If it is desirable
to put the net in a valise or trunk when one is travelling,
the handle may be made in two pieces by sawing it in
two in the middle, having a close-fitting tin or brass
ferrule made to hold it together when in use.

Chloroform has been used to some extent for killing
insects, but what is called a "poison-bottle" is preferable to this, on account both of expense and of ease
in use. This is made by placing in a large-mouthed
bottle several pieces of cyanide of potassium, the amonnt
depending upon the size of the bottle. If the bottle
is large and the glass thin, it is better to break the cyanide into pieces not larger than a pea, as otherwise the
bottle may be broken by expansion of the poison-cake.
After the chemical is in the bottle, pour in water to the
depth of half an inch or less, and slowly sprinkle in
plaster of Paris till a hard, dry cake is formed, having
some loose plaster on top of the cake. Upon turning the
side and rolling it round, this will absorb any moisture
on the inside of the bottle. Wipe down the sides now
with a cloth, using a stick if necessary, pour out the
dry plaster, wipe again both inside and out, put in the
cork, and the bottle is ready for use. Quinine-bottles
are a very good size for small insects. The glass jars
with tin tops in which "Old Reliable" baking-powder
is put up make excellent bottles for general use. It is
better to have several poison-bottles, so that one may be
had for use at any time without disturbing those that

may be in another waiting to be spread. In excursions for butterflies it is well to have two or more bottles, so that when an insect is killed it may be put into another bottle, and not be beaten by the fluttering wings of the next capture. Even then but few should be kept together in the second bottle, as they soon get rubbed by being carried about. Some means of pinning them in the field as soon as they have been in the poison-bottle long enough to insure their not coming to life again is preferable. To avoid this rubbing, the writer obtained a small tin box with a handle at the top, and the lid fastening with a clasp, made a · cyanide-cake in the bottom, and put a sheet of cork round the inside, with the edge coming just to the top of the box. This box is eight and a half inches long by six wide, and five high ; but that is rather small to hold a large number of captures. With a box of this kind it is necessary to use only one bottle, as when the insects are pinned on the inside of the box they are still under the influence of the poison, and hence may be pinned as soon as they become quiet in the bottle. This has another advantage over a box without the cyanide, as the specimens need not be spread till the next day after they are captured, or even longer. If allowed to remain long in the box, however, the pins are liable to corrode.

The subject of using the net may be passed over briefly, as a little practice is of more value than pages of verbal directions. When the insect is in the net, a quick turn of the hand brings the top down with a fold in the bag and prevents its escape. Then by carefully getting its body between the thumb and finger outside the net, with the wings closed back to back, fluttering is pre-

vented, and the wings are kept from being broken or
the scales from being rubbed from the body and wings.
After this is done, the poison-bottle, with the cork out,
may be inserted under the net and the butterfly let into it,
where he will soon succumb to the poisonous fumes.   In
taking small specimens out of the net it is not necessary
to seize them between the thumb and finger : with the
hoop on the ground, the bottom of the net may be raised
up, when they will fly upward as far as they can get.
Inserting now the open bottle into the net, the specimens
are easily secured.

After the butterflies are captured, what we shall do with
them depends upon whether they are to be at once prepared for the cabinet, or whether for any reason this
cannot be done.   Only entomological pins, or those prepared specially for this use, should be used in pinning
specimens.   For butterflies, Nos. 3, 4, 5, and 6 of Klaeger's
make are considered the best, suiting the pin to the size

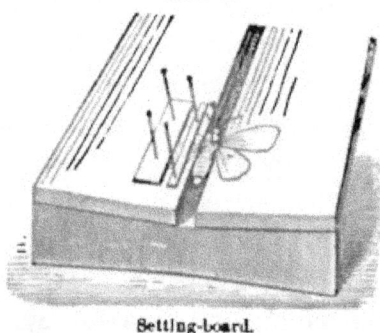

FIG. 9.

Setting-board.

of the insect.   The pin
should be inserted into
the middle of the thorax,
and passed through till
at least one-fourth of
the pin is above the
body, some preferring
as much as one-third
being left above.   This
will give room to take
hold of the pin in trans-
ferring from box to box without injury to the covering
of the thorax, and will bring the specimens to the same
height in the cabinet.   After pinning, the wings should

be spread in the manner represented by Fig. 9, on what is called a setting-board. This may be made of any length desired, and several sizes should be on hand to accommodate different-sized specimens. They may be made by taking clapboards or siding, sawing them into strips, and nailing them to blocks of wood one inch high, as in the figure, the thin edge of the board inward, with a space between for the bodies of the insects, varying according to the size of the specimen to be pinned on the board. This makes the boards slope a little towards the middle, and brings the outer part of the wings a little higher than next to the body; but this is best, as when taken from the boards they may droop a little. If the setting-boards are twenty-three inches long, it will be necessary to support them by a block in the middle. Under the space between the boards should be fastened a narrow strip of one-eighth inch cork, or a piece of thin pasteboard, through which the pins must be pushed till the lower side of the wings, when spread, comes on a level with the boards.

Fig. 10.

Setting-needle.

In spreading insects' wings setting-needles (see Fig. 10) should be used. In handling specimens a pair of spring forceps with smooth points are essential to prevent rubbing by the fingers. The setting-needle is made by taking a medium-sized needle in a pair of pliers and forcing the eye end into a piece of soft wood. Five of these will be found convenient,—one with which to bring the wings down if they stand erect, and the other four to bring the wings round in place, inserting each one into the soft setting-board through the wing when the latter

4*

is where it is wanted. It is customary now among entomologists to bring the fore wings forward until the hind
margins of these wings shall form a straight line, as in
Fig. 32, and then bring the hind wings far enough round
to look natural. When the wings are in place, put
on each side from one to two narrow strips of paper, as
shown in the figure.

After the insects are spread on the boards they may
be put into a drying-case, where they should remain from
five to ten days, according to the size of the specimens
and the state of the weather. A convenient case may
be made in the form of a box long enough to hold the
boards, set on edge, with shelves put in it three inches
apart, and with a door in front. It may be deep enough
for two of the boards to go on each shelf. If the back
of the case is made of wire-cloth the specimens will dry
more readily.

If conveniences are not at hand for spreading butterflies when caught, as in travelling, they may be pinned,
but not spread, and put into empty boxes, to be relaxed
and spread at some future time; or they may be put into
papers or small envelopes, with such notes as to place and
date of capture, etc., as may be of interest marked on the
outside. To prepare a paper for this purpose, take a strip
of ordinary writing-paper a little longer than wide, and
fold it obliquely across the middle so that a quarter of an
inch shall project beyond each of the sides of the triangle
thus made. With the butterfly inside of this, the wings
folded back to back, and the projecting part folded
over the edge on each side, a receptacle is formed which
will keep the insect in good condition as long as desired. The size of the papers should vary with the size

of the specimen to be put up. Insects put up in this way may be packed in boxes and sent through the mails to any distance with little danger of injury.

When desirable to prepare specimens, not spread, for the cabinet, they may be put into a jar or box having two inches of wet sand in the bottom, over which a couple of thicknesses of paper have been placed. By remaining in such a place from twenty-four to forty-eight hours, or longer if not pliable by that time, the specimens become softened so that the wings may be spread the same as fresh specimens. The jar or box containing the specimens should be kept in a cool place, as otherwise the insects may mould before they are relaxed enough to be spread.

Some form of a cabinet to hold the specimens after they are dry is a necessity. The best form is that consisting of a series of closed drawers, all enclosed by doors, as this double enclosing insures partial immunity from museum pests. Among the many patterns or styles the simple is often as good as the more complicated. This may consist of drawers of any desired size, with a glass top set into a frame that matches tightly on to the lower part. For a large cabinet the glass may be sixteen by twenty inches, and the drawers one and three-fourths or one and seven-eighths inches deep on the inside from the bottom to the glass. As many as seventy-two of these drawers may be arranged in a cabinet in three upright rows, or a less number if desired. For a smaller cabinet the drawers may be smaller, as, for instance, fourteen by sixteen inches glass with the same depth ; and any number from sixteen to forty, arranged in two rows. It is better to have the cabinet made of some hard wood, as cherry or black walnut, though the drawers, all but the front, may be of pine. When made, the drawers should be lined on the bottom with insect cork, and this and the sides, up to the glass, covered with white paper. If the paper be ruled both ways with blue lines one-fourth of an inch apart, this will facilitate putting specimens in with regularity, and will not detract from the looks.

Some use boxes made in the form of books, which are convenient on some accounts, but cannot be recommended except as a temporary expedient.

Museum pests are great destroyers of collections. These consist of one or more species of Dermestidæ, a family of small beetles, which in the larva state eat any dead animal matter if it be dry, and one or more species of small mites. Camphor gum wrapped in a piece of thin cloth and put into a corner of the drawer, or a naphthaline cone pinned in, will destroy the mites, but the Dermestes must be sought and killed. Nothing short of vigilance will keep them out. When a cabinet is free from them, careful guarding against their introduction in new specimens added to it will generally insure immunity from them. Where specimens are suspected of being infested, they should be placed in a box away from the cabinet and watched, and not introduced until known to be free from these pests.

One of the best means of obtaining good specimens of many butterflies is by raising them from the eggs or captured larvæ. Besides good specimens, a knowledge of the preparatory stages is thus obtained, and this is not less important than the habits and other items we learn about the imagines. It is now well known, chiefly through the investigations of Mr. Edwards, that the

females of most, if not of all, of the species of butterflies when caught and confined with the proper food-
plant will readily deposit eggs. If the plant be small,
it may be transferred to a flower-pot, and the whole
covered with a Swiss muslin bag, supported by two or
three sticks set in the dirt in the pot. If the food-plant
is a tree or bush, then a limb may be enclosed in the
bag with the butterfly in it. The female placed in the
bag may not deposit eggs at once, but in most cases she
will do so within two or three days. Some species do
not lay their eggs till some time after their emergence
from the chrysalis, as the eggs are not sufficiently
matured in the ovaries. With such species it is better
to take a specimen which by its worn appearance shows
that it has been some time from the chrysalis. It may
be desirable to keep the butterfly alive for several days,
and this can be done only by feeding her, as without
food she would starve. A method recommended by
Mr. Edwards is to put raw dried apples into a small
dish with a little sugar and water. The insect will eat
this readily, and by this means eggs may be secured
when they would not be otherwise.

If the food-plant is not known, several plants may be
tried till one is found upon which the butterfly will
oviposit. Often a food-plant may be guessed by know-
ing what an allied species feeds upon; though this does
not always hold true. In the part of this work devoted
to the descriptions of species, the food-plants of the
species are given so far as known. As will be seen, there
are a number of species of whose preparatory stages
nothing is known, embracing, among others, nearly all
of the Hesperidæ.

In feeding the larvæ, some entomologists leave the muslin bag over the limb where the eggs were deposited until the eggs hatch, and allow the larvæ to feed on the leaves thus enclosed. This is a good way to obtain chrysalides for butterflies with little expense of labor in feeding and caring for the larvæ, if only the perfect insects are desired, but it gives very little chance for observations.

If it is desirable to study the larvæ as they grow (and nothing about entomology can be more interesting), the eggs, with the leaves on which they are placed, should be removed to some vessel where they will not dry up, and where they will be secure from the attacks of ants. Ordinary jelly-glasses with tin tops are good for this purpose. When the eggs begin to hatch, the young larvæ may be transferred to another jelly-glass, or, if there are only a few of them, they may be left for a day or two in the same dish, introducing tender leaves for them to eat. Mr. Edwards suggests the use of tubes for a few of the larvæ, so that they may be more easily observed while young than in a larger dish. The time of depositing the eggs should be noted, also the time of hatching, and the shape and color of the young larvæ, making observations every twelve or twenty-four hours.

The larvæ of butterflies moult or shed their skins from four to five times in coming to maturity, and usually they present different colors and markings each time they change their outer covering. These notes, taken at least as often as they moult, and a description of the egg and chrysalis, with the dates at which these changes take place, form the life-history of the species. To this should be added any other items of interest that may be observed, such as the food-plant, whether solitary or gregarious, the manner of feeding and place of resting, place and manner in which the chrysalis is formed, manner of depositing the eggs, etc. Such life-histories form a valuable part of our knowledge of these insects.

Moulting is a critical period in the history of a caterpillar. To make preparation for it the larva fixes its feet to some object, and after remaining quiet for a time bursts the skin open on the back of the thorax, and through this opening comes out of its old skin. At first all the external parts are soft and tender, for which reason it is easily injured. Soon the parts harden by evaporation of the water, and the caterpillar then resumes its feeding. During the time of moulting it should not be transferred from one vessel to another if it can be avoided. The time of moulting may be known by the enlargement of the neck or second segment, caused by the partial withdrawing of the head from the old skin and pushing it up against this part.

If the larvæ when first hatched are placed in tubes, they may afterwards be placed in a jelly-dish, where they may be kept till of considerable size. The writer has kept several species in jelly-dishes till they were three-fourths of an inch long. From this they may be transferred to the breeding-cage, or vivarium. Some entomologists use a cage made of wire gauze over a framework, with a zinc bottom to hold dirt and the food, but these are expensive. The writer has used for a number of years boxes of different sizes with a glass in front for a door, sliding in upright grooves. Upon the bottom of the box is placed two and a half inches of garden-soil and sand,—just enough of the latter to prevent the soil from drying in hard cakes. By wetting the dirt occasionally,

and placing the food-plant on the dirt or setting it up
against the sides, the conditions can be made very nearly
what they are where the larvæ feed unconfined on the
grass or tree in their freedom; and this making their
condition as nearly as possible what it is in nature is
essential to successful breeding.

It is a mistaken idea that larvæ will not bear confinement in close vessels. Being obliged to leave home
for a time when two species were about half grown, I
had some of each put into jelly-dishes with their food-
plants and took them with me. They were kept in
these dishes till they reached maturity by being fed each
day, and they produced the imagines without the loss of
a single individual; and this is not the only instance
in which the writer has kept them in this way.

In an article in the "Canadian Entomologist" on
rearing larvæ (vol. xvi. page 116), Mr. W. H. Edwards
says, " Larvæ bear confinement in tight glasses well, and
I often receive them from correspondents as distant as
Florida or California, through the mails, in good condition. The plants keep well in this sort of confinement
also. I have never used what are known as breeding-
cages, which are expensive if purchased, and are trouble-
some to make at home." Farther on in the same article
he says, "As for large larvæ, as of the Papilios, I
generally use powder-kegs (wood), or nail-kegs, one or
the other of which can be had anywhere. Remove the
top hoop, and use the second one to bind down the
cotton-cloth cover; put a little earth in the bottom, and
in it set a two-quart glass fruit-jar filled with water, in
which branches of the food-plant are placed. No
further care is required than to substitute fresh branches

for the old ones as the leaves are consumed, and the larvæ will go on to pupation. This is when it is desired to get chrysalides by wholesale.

"But where observation of the larvæ is necessary, glass tubes and tumblers, and gauze-covered flower-pots, and tin pails and nail-kegs, will answer all the purposes of elaborate and expensive cages, and be more satisfactory, I apprehend. At any rate, all my work is done in this way." A reference to Mr. Edwards's publications will attest the success of his methods of rearing butterfly larvæ.

USE OF THE KEY.

In the preparation of the following key for the aid of the student in identifying specimens, an effort has been made so to combine a key to the genera with one to the species that they shall be one key, and at the same time have the merits of separate keys. This has been done by inserting, at the end of a description referring to a genus, the generic name as well as the number referring to where the specific descriptions begin. If the genus to which a given specimen belongs is known, it will not be necessary to begin at the first of the key and go through till the particular genus is reached, but by running the eye over the names at the right-hand side of the page the genus may be found, and the number after the generic name will direct where to go to find the species.

In the preparation of the key such characters have been used as would be the most readily recognized by the student, facility in identification being aimed at, though by so doing the tables were made in a measure artificial. The numbers in parentheses after the *species*

c   d   5

refer to the numbers at the head of each specific description in the body of the work, the other numbers at the *right-hand side of the page* refer to other parts of the key.

NAMES OF BUTTERFLIES.

Few of our American butterflies have common names that are generally recognized. The few that have common names are mostly such as have obtained them by reason of their being injurious insects, such as the Rape or European Cabbage Butterfly, etc. For this reason only the scientific names are used in this work. If it is desirable to use a common name, the specific name can be used as such; indeed, this has for some time been the custom in many localities. For instance, *Papilio Asterias* is spoken of as the Asterias butterfly, *P. Ajax* as the Ajax butterfly, etc. *Pieris Rapæ*, because of its being brought to this country from Europe, is usually called the European Cabbage butterfly, though it is often called the Rape butterfly.

The scientific names are, like the scientific names of other groups of natural objects, Latin words, and as such are subject to the rules of that language in pronunciation. By observing the place of accent as given in the following list, and remembering that, with few exceptions, in Latin each vowel makes a syllable, little difficulty need be experienced in pronunciation. To those who are classical scholars no such suggestions are necessary.

The scientific name consists of two words, the first the generic name, or the name of the genus. This has nearly its parallel in the last word of a man's name. The second is the specific name, or that which is used to identify the particular kind. In the names of human beings, the given name, or that which is used to designate a particular individual, in a measure corresponds with this. After the specific name is given, in an abbreviated form, the name of the entomologist who described the insect originally under the specific name here used.

ACCENTUATED LIST OF THE BUTTERFLIES OF THE EASTERN UNITED STATES.

1. Pa-pil′i-o A′jax, Linnæus.
2. Pa-pil′i-o Phi-le′nor, Linnæus.
3. Pa-pil′i-o As-te′ri-as, Fabricius.
   "   "   aberr. Cal-ver-ley′i-i, Grote.
   "   "   var. As-te-ro-i′des, Reakirt.
4. Pa-pil′i-o Tro′i-lus, Linnæus.
5. Pa-pil′i-o Pal-a-me′des, Drury.
6. Pa-pil′i-o Tur′nus, Linnæus.
   "   "   dim. form Glau′cus, Linnæus.
7. Pa-pil′i-o Cres-phon′tes, Cramer.
8. Pa-pil′i-o Po-lyd′a-mas, Linnæus.
9. Pi′e-ris Il-a-i′re, Godart.
10. Pi′e-ris Mo-nus′te, Linnæus.
11. Pi′e-ris Pro-tod′i-ce, Boisduval and Leconte.
    "   "   form Ver-na′lis, Edwards.
12. Pi′e-ris Na′pi, Esper.
    "   "   aberr. Vir-gin-i-en′sis, Edwards.
    "   "   form Ol-e-ra′ce-a Æs′ti-va, Harris
13. Pi′e-ris Vir-gin-i-en′sis, Edwards.
14. Pi′e-ris Ra′pæ, Linnæus.
    "   "   var. Man′ni, Mayer.
    "   "   var. No′væ An′gli-æ, Scudder.
15. Na-tha′lis I′o-le, Boisduval.
    "   "   var. I-re′ne, Fitch.
16. An-thoch′a-ris O-lym′pi-a, Edwards.
17. An-thoch′a-ris Ge-nu′ti-a, Fabricius.
18. Cal-lid′ry-as Eu-bu′le, Linnæus.
19. Cal-lid′ry-as Sen′næ, Linnæus.

20. Cal-lid′ry-as Phil′e-a, **Linnæus.**
21. Cal-lid′ry-as A-gar′i-the, **Boisduval.**
22. Kric-o-go′ni-a Lys′i-de, **Godart.**
　　　　　　　　　" 　　　　　" 　　　　form Te-ris′sa, **Lucas.**
23. Co′li-as Cœ-so′ni-a, Stoll.
24. Co′li-as Eu-ryth′e-me, Boisduval.
　　　　" 　　　　　" 　　　form A-ri-ad′ne, Edwards.
　　　　" 　　　　　" 　　　" 　Kee-way′din, Edwards.
25. Co′li-as Phi-lod′i-ce, Godart.
26. Co′li-as In-te′ri-or, **Scudder.**
27. Te′ri-as Ni-cip′pe, **Cramer.**
28. Te′ri-as Mex-i-ca′na, **Boisduval.**
29. Te′ri-as Li′sa, **Boisduval and Leconte.**
30. Te′ri-as De′li-a, **Cramer.**
31. Te′ri-as Ju-cun′da, **Boisduval and Leconte.**
32. Hel-i-co′ni-a Char-i-to′ni-a, **Linnæus.**
33. Dan′a-is Ar-chip′pus, **Fabricius.**
34. Dan′a-is Ber-e-ni′ce, **Cramer.**
35. Co-læ′nis Ju′li-a, **Fabricius.**
36. A-grau′lis Va-nil′læ, **Linnæus.**
37. Ar-gyn′nis I-da′li-a, **Drury.**
　　　　　" 　　　　　" 　　　aberr. Ash′ta-roth, **Fisher.**
38. Ar-gyn′nis Di-a′na, **Cramer.**
39. Ar-gyn′nis Cyb′e-le, **Fabricius.**
40. Ar-gyn′nis Aph-ro-di′te, **Fabricius**
41. Ar-gyn′nis Al-ces′tis, **Edwards.**
42. Ar-gyn′nis At-lan′tis, **Edwards.**
43. Ar-gyn′nis My-ri′na, **Cramer.**
44. Ar-gyn′nis Mon-ti′nus, Scudder.
45. Ar-gyn′nis Bel-lo′na, **Fabricius.**
46. Eup-toi-e′ta Clau′di-a, Cramer.
47. Mel-i-tæ′a Pha′e-ton, Drury.
　　　　" 　　　　　" 　　　aberr. Su-per′ba, **Strecker.**
　　　　" 　　　　　" 　　　" 　Phæ-thu′sa, **Hulst.**
48. Mel-i-tæ′a Har-ris′i-i, **Scudder.**
49. Phy-ci-o′des Nyc′te-is, **Doubleday and Hewitson.**
50. Phy-ci-o′des Car-lo′ta, Reakirt.
51. Phy-ci-o′des **Pha′on,** Edwards.
52. Phy-ci-o′des Tha′ros, Drury.
　　　　" 　　　　　" 　　　form Mar′ci-a, **Edwards.**

Phy-ci-o'des Tha'ros, form Mor'phe-us, Fabricius.
" " " aberr. Pack-ar'di-i, Saunders.
53. Phy-ci-o'des Bates'i-i, Reakirt.
54. E-re'si-a Fris'i-a, Poey.
55. Grap'ta In-ter-ro-ga-ti-o'nis, Fabricius.
" form Fa-briç'i-i, Edwards.
" " Um-bro'sa, Lintner.
56. Grap'ta Com'ma, Harris.
" " form Har-ris'i-i, Edwards.
" " " Dry'as, Edwards.
57. Grap'ta Fau'nus, Edwards.
58. Grap'ta Graç'i-lis, Grote and Robinson.
59. Grap'ta Prog'ne, Cramer.
60. Grap'ta J Al'bum, Boisduval and Leconte.
61. Va-nes'sa An-ti'o-pa, Linnæus.
" " aberr. Lint-ne'ri-i, Fitch.
62. Va-nes'sa Mil-ber'ti-i, Godart.
63. Py-ra-me'is At-a-lan'ta, Linnæus.
64. Py-ra-me'is Hun'te-ra, Fabricius.
65. Py-ra-me'is Car'du-i, Linnæus.
66. Ju-no'ni-a Cœ'ni-a, Hübner.
67. A-nar'ti-a Jat'ro-phæ, Linnæus.
68. Eu'ni-ca Mon'i-ma, Cramer.
69. Ti-me'tes Pet're-us, Cramer.
70. Vic-to-ri'na Sten'e-les, Linnæus.
71. Di-a-de'ma Mi-sip'pus, Linnæus.
72. Li-men-i'tis Ur'su-la, Fabricius.
73. Li-men-i'tis Ar'the-mis, Drury.
" " form Lam'i-na, Fabricius.
" " " Pro-ser'pi-na, Edwards.
74. Li-men-i'tis Di-sip'pus, Godart.
" " var. Flor-i-den'sis, Strecker.
" " aberr. Pseu-do-do-rip'pus, Strecker
75. Li-men-i'tis E'ros, Edwards.
76. Ap-a-tu'ra Cel'tis, Boisduval and Leconte.
77. Ap-a-tu'ra A-liç'i-a, Edwards.
78. Ap-a-tu'ra Cly'ton, Boisduval and Leconte.
" " form Pro-ser'pi-na, Scudder.
" " " O-cel-la'ta, Edwards.
79. Ap-a-tu'ra Flo'ra, Edwards.

80. Pa'phi-a Trog-lod'y-ta, Fabricius.
81. De'bis Port-lan'di-a, Fabricius.
82. Ne-o-nym'pha Can'thus, Boisduval and Leconte.
83. Ne-o-nym'pha Gem'ma, Hübner.
84. Ne-o-nym'pha A-re-o-la'tus, Smith and Abbott.
85. Ne-o-nym'pha Eu'ry-tris, Fabricius.
86. Ne-o-nym'pha So-syb'i-us, Fabricius.
87. Sat'y-rus Pe-ga'la, Fabricius.
88. Sat'y-rus Al'o-pe, Fabricius.
    "   "   form Ma-rit'i-ma, Edwards.
    "   "   " Neph'e-le, Kirby.
    "   "   " O-lym'pus, Edwards.
89. Chi-o-no'bas Jut'ta, Hübner.
90. Chi-o-no'bas Se-mid'e-a, Say.
91. Lib-y-the'a Bach-man'ni, Kirtland.
92. Ca-leph'e-lis Cæ'ni-us, Linnæus.
93. Ca-leph'e-lis Bo-re-a'lis, Grote and Robinson
94. Eu-me'ni-a At'a-la, Poey.
95. Thec'la Ha-le'sus, Cramer.
96. Thec'la M Al'bum, Boisduval and Leconte.
97. Thec'la Fa-vo'ni-us, Smith and Abbott.
98. Thec'la Au-tol'y-cus, Edwards.
99. Thec'la Hu'mu-li, Harris.
100. Thec'la A-ca'di-ca, Edwards.
101. Thec'la Ed-wards'i-i, Saunders.
102. Thec'la Witt-feld'i-i, Edwards.
103. Thec'la Cal'a-nus, Hübner.
    "   "   var. Lo-ra'ta, Grote and Robinson.
104. Thec'la On-ta'ri-o, Edwards.
105. Thec'la Stri-go'sa, Harris.
106. Thec'la Smi-la'cis, Boisduval and Leconte.
107. Thec'la A'cis, Drury.
108. Thec'la Po'e-as, Hübner.
109. Thec'la Col-u-mel'la, Fabricius.
110. Thec'la Au-gus'tus, Kirby.
111. Thec'la I'rus, Godart.
    "   "   var. Ar'sa-ce, Boisduval and Leconte.
112. Thec'la Hen'ri-ci, Grote and Robinson.
113. Thec'la Ni'phon, Hübner.
114. Thec'la Læ'ta, Edwards.

115 Thec'la Ti'tus, Fabricius.
116. Fen-i-se'ca Tar-quin'i-us, Fabricius.
117. Chrys-o-pha'nus Di-o'ne, Scudder.
118. Chrys-o-pha'nus Tho'e, Boisduval and Leconte.
119. Chrys-o-pha'nus Ep-ix-an'the, Boisduval and Leconte.
120. Chrys-o-pha'nus Hy-poph'le-as, Boisduval.
121. Ly-cæ'na Lyg'da-mus, Doubleday.
122. Ly-cæ'na Scud-de'ri-i, Edwards.
123. Ly-cæ'na Pseu-dar-gi'o-lus, Boisduval and Leconte.
          "   "   form Lu'ci-a, Kirby.
          "   "   " Mar-gi-na'ta, Edwards.
          "   "   " Vi-o-la'ce-a, Edwards.
          "   "   " Ni'gra, Edwards.
          "   "   " Neg-lec'ta, Edwards.
124. Ly-cæ'na Co-myn'tas, Godart.
125. Ly-cæ'na Fi-le'nus, Poey.
126. Ly-cæ'na I-soph-thal'ma, Herrick-Schæffer.
127. Ly-cæ'na Ex'i-lis, Boisduval.
128. Ly-cæ'na Am'mon, Lucas.
129. Ly-cæ'na The-o'nus, Lucas.
130. Car-te-ro-ceph'a-lus Man'dan, Edwards.
131. Car-te-ro-ceph'a-lus O'ma-ha, Edwards.
132. An-cy-lox'y-pha Nu'mi-tor, Fabricius.
133. Thy-mel'i-cus Pow'e-shiek, Parker.
134. Pam'phi-la Mas-sa-so'it, Scudder.
135. Pam'phi-la Zab'u-lon, Boisduval and Leconte.
          "   "   var. Ho-bo'mok, Harris.
          "   "   form Po-ca-hon'tas, Scudder.
          "   "   var. Quad-ra-qui'na, Scudder.
136. Pam'phi-la Sas'sa-cus, Harris.
137. Pam'phi-la Me'te-a, Scudder.
138. Pam'phi-la Un'cas, Edwards.
139. Pam'phi-la Sem-i-no'le, Scudder.
140. Pam'phi-la Le-o-nar'dus, Harris.
141. Pam'phi-la Mes'ke-i, Edwards.
142. Pam'phi-la Hu'ron, Edwards.
143. Pam'phi-la Phy-læ'us, Drury.
144. Pam'phi-la Bret'tus, Boisduval and Leconte.
145. Pam'phi-la O'tho, Smith and Abbott.
          "   "   var. E-ger'e-met, Scudder.

146. Pam'phi-la Peck'i-us, Kirby.
147. Pam'phi-la Mys'tic, Scudder.
148. Pam'phi-la Cer'nes, Boisduval and Leconte.
149. Pam'phi-la My'us, French.
150. Pam'phi-la Man-a-ta'a-qua, Scudder.
151. Pam'phi-la Ver'na, Edwards.
152. Pam'phi-la Ves'tris, Boisduval.
153. Pam'phi-la Met-a-com'et, Harris.
154. Pam'phi-la Ac'ci-us, Smith and Abbott.
155. Pam'phi-la Lo'am-mi, Whitney.
156. Pam'phi-la Mac-u-la'ta, Edwards.
157. Pam'phi-la Pa-no'quin, Scudder.
158. Pam'phi-la O-co'la, Edwards.
159. Pam'phi-la Eth'li-us, Cramer.
160. Pam'phi-la Bi-mac'u-la, Grote and Robinson.
161. Pam'phi-la Pon'ti-ac, Edwards.
162. Pam'phi-la Di'on, Edwards.
163. Pam'phi-la Ar'pa, Boisduval and Leconte.
164. Pam'phi-la Pa-lat'ka, Edwards.
165. Pam'phi-la Vi-tel'li-us, Smith and Abbott.
166. Pam'phi-la Del-a-wa're, Edwards.
167. Pam'phi-la Bys'sus, Edwards.
168. Pam'phi-la O-sy'ka, Edwards.
169. Pam'phi-la Eu-fa'la, Edwards.
170. Pam'phi-la Fus'ca, Grote and Robinson.
171. Pam'phi-la Hi-an'na, Scudder.
172. Pam'phi-la Vi-a'tor, Edwards.
173. Am-bly-scir'tes Vi-a'lis, Edwards.
174. Am-bly-scir'tes E'os, Edwards.
175. Am-bly-scir'tes Sam'o-set, Scudder.
176. Am-bly-scir'tes Tex'tor, Hübner.
177. Pyr'gus Tes-sel-la'ta, Scudder.
178. Pyr'gus Cen-tau're-æ, Rambur.
179. Nis-o-ni'a-des Bri'zo, Boisduval and Leconte.
180. Nis-o-ni'a-des Ic'e-lus, Lintner.
181. Nis-o-ni'a-des Som'nus, Lintner.
182. Nis-o-ni'a-des Lu-cil'i-us, Lintner.
183. Nis-o-ni'a-des Per'si-us, Scudder.
184. Nis-o-ni'a-des Au-so'ni-us, Lintner.
185. Nis-o-ni'a-des Mar-ti-a'lis, Scudder.

186. Nis-o-ni'a-des Ju-ve-na'lis, Fabricius.
187. Nis-o-ni'a-des Pe-tro'ni-us, Lintner.
188. Nis-o-ni'a-des Næ'vi-us, Lintner.
189. Phol-i-so'ra Ca-tul'lus, Fabricius.
190. Phol-i-so'ra Hay-hurst'i-i, Edwards:
191. Eu-da'mus Pyl'a-des, Scudder.
192. Eu-da'mus Ba-thyl'lus, Smith and Abbott.
193. Eu-da'mus Lyç'i-das, Smith and Abbott.
194. Eu-da'mus Cel'lus, Boisduval and Leconte.
195. Eu-da'mus Zes'tos, Hübner.
196. Eu-da'mus Tit'y-rus, Fabricius.
197. Eu-da'mus Pro'te-us, Linnæus.
198. E-ryç'i-des Bat-a-ba-no', Lefebvre.
199. E-ryç'i-des A-myn'tas, Fabricius.
200. Meg-a-thy'mus Yuc'cæ, Boisduval and Leconte.
201. Meg-a-thy'mus Cof-a-qui', Strecker.

ANALYTICAL KEY FOR THE DETERMINATION OF THE BUTTERFLIES OF THE EASTERN UNITED STATES.

1.

Antennæ filiform, terminating in a knob or club,
                                                    BUTTERFLIES,   2
Antennæ not knobbed,   MOTHS.

2.

Having six feet adapted for walking,   3
Having four feet adapted for walking, the front pair aborted,   35

3.

Body rather slender, width of thorax from one-eighth to one-
    sixth the length of hind margin of fore wings,   4
Body robust, width of thorax from one-fourth to one-half the
    length of hind margin of fore wings,   Hesperidæ, 120

4.

General or ground colors black, white, or yellow; size generally
    from medium to large,   Papilionidæ,   5
Colors blue, coppery, or blackish; size generally small,
                               Lycænidæ, 89

Colors black and green, with fulvous abdomen; or brownish
fulvous, with many black spots,   **Erycinidæ, ♀s, 87**
Colors black and fulvous, palpi beak-like,

Libythea Bachmanni, ♀ (91)

5.

Large species, hind wings tailed; or black **with** submarginal
bands of yellow spots,   Papilio,   6
Small or medium-sized species, hind wings not tailed, Pierinæ, 14

6.

Ground or principal color black,   7
Ground color yellow, with several black bands, **Papilio Turnus,** (6)

7.

Wings crossed by a number of greenish or greenish-yellow
bands, red at anal angle,   **Papilio Ajax,** (1)
Wings crossed by one or two rows of yellow or greenish spots,   8

8.

Wings crossed by one row of spots,   9
Wings crossed by two rows of spots,   11

9.

Wings tailed,   10
Wings not tailed,   **Papilio Polydamas,** (8)

10.

No blue clouds on hind wings, spots greenish, **Papilio Philenor,** (2)
Blue clouds on hind wings, spots yellow,

Papilio Turnus, var. Glauca, ♀ (6)

11.

Rows of spots yellow,   12
Rows of spots bluish or greenish, two rows on hind wings and
near the posterior angle of fore wings, **Papilio Troilus,** (4)

12.

Spots parallel with the outer margin,   13
Rows of spots not parallel with the outer margin,

Papilio Cresphontes, (7)

18.

Orange anal patch pupilled with black,   **Papilio Asterias,** (3)

Orange anal patch not pupilled with black,
Papilio Palamedes, (5)
14.
Antennæ abruptly terminating in an ovoid club,   15
Antennæ truncated at the end and obconic, or terminating insensibly in an obconic club,   22
15.
Abdomen shorter than the hind wings; color white or very
pale yellow,   16
Abdomen as long as or longer than the hind wings; color
lemon-yellow, black at apex, along hind margin of fore
wings and costa of hind wings,   Nathalis Iole, (15)

16.
Under side of hind wings without marks, or marked along the
veins,   Pieris,   17
Under side of hind wings covered with a greenish net-work,
Anthocharis,   21
17.
Fore wings with no black bar at the end of the cell,   18
Fore wings with a black bar at the end of the cell, and more or
less of a black border,   Pieris Protodice, (11)

18.
Under side of hind wings plain white, apex of fore wings
black,   Pieris Ilaire, (9)
Under side of hind wings plain yellow,   Pieris Rapæ, (14)
Under side of hind wings marked along the veins,   19

19.
Black border to both wings,   Pieris Monuste, (10)
No black border to the wings,   20

20.
Under side of hind wings pale yellow, with brownish along
the veins,   Pieris Napi, aberr. Virginiensis, (12)
Under side of hind wings white, with brownish along the veins,
Pieris Virginiensis, (13)
21.
No orange at apex of fore wings,
Anthocharis Olympia, (16)
A large orange apical patch,   Anthocharis Genutia, (17)

22.

Antennæ insensibly terminating in a club, truncate at the end,

Callidryas, 23

Antennæ terminating in an obconic club,   26

23.

Color above, lemon-yellow or dirty yellowish white; beneath,
  lemon-yellow, with or without orange tint,   24
Color above, orange or whitish orange,   25

24.

Clear lemon-yellow above and below, with slight greenish tint;
  female with discal dot and terminal row of spots,

Callidryas Eubule, (18)

Color above, lemon-yellow, or dirty whitish yellow; beneath,
  orange-tinted; female with discal dot on fore wings and
  terminal border,   Callidryas Sennæ, (19)

25.

Color light orange with reddish-orange spots,

Callidryas Philea, (20)

Color light orange or orange-tinted white; orange-tinted beneath; subterminal band on under side of fore wings
  straight,   Callidryas Agarithe, (21)

26.

Hind wings with an orange discal spot; both wings with black
  border,   Colias, 27
Hind wings without discal spot,   30

27.

Ground color yellow,   28
Ground color orange, or at least an orange patch in the middle
  of fore wings,   Colias Eurytheme, (24)

28.

Fore wings black at base, the yellow making a "dog's head,"
  with the discal dot for an eye,   Colias Cæsonia, (23)
Fore wings without black at base,   29

29.

Under side with a submarginal row of dark points,

Colias Philodice, (25)

Under side without submarginal row of dark points,

Colias Interior, (26)

30.

Color yellow or orange, with black terminal border,   Terias, 31
Color white, base of fore wings yellow,   Kricogonia Lyside, (22)

31.

Hind wings with a prominent angle, color yellow,

Terias Mexicana, (28)
Hind wings rounded, not angled,   32

32.

Ground color orange,   Terias Nicippe, (27)
Ground color yellow,   33

33.

Fore wings without black on the hind margin,   Terias Lisa, (29)
Fore wings with black along the hind margin,   34

34.

Under side with pink at apex of fore wings and over hind
    wings,   Terias Delia, (30)
Under side without pink,   Terias Jucunda, (31)

35.

Small species, brownish fulvous with many black spots; or
    black and green, with abdomen orange,   Erycinidæ, ♂s, 87
Not having the above characters,   Nymphalidæ, 36

36.

Fore wings long and narrow, twice as long as wide,   37
Fore wings not long and narrow,   38

37.

Black, with yellow zebra stripes,   Heliconia Charitonia, (32)
Fulvous, with black border and oblique stripe, Colænis Julia, (35)

38.

Palpi remote, not extending much beyond the head; discal cell
    of hind wings closed; a black spot on vein of hind wings
    of ♂,   Danais, 39
Palpi nearly connivent, porrect, discal cell generally open;
    veins of fore wings not dilated at base,   40

Palpi close, elevated, **very hairy**; discal cell **always closed**;
veins of fore wings **usually dilated at base**,   79
Palpi several times as long **as the head, contiguous, in the form**
of a beak; wings angled; **females six-footed**,

Libythea Bachmanni, ♂(91)

39.

Color fulvous, veins black,   Danais Archippus, (33)
Color fulvous brown, veins not black,   Danais Berenice, (34)

40.

Eyes **naked**,   41
Eyes **hairy**,   69

41.

Club of antennæ **short, ovoid, usually flattened in dry specimens**,   42
Club of antennæ **obconic or gradually terminating in a knob**,   62

42.

Outer margin of fore wings **sinuous**,   43
Outer margin of fore wings **not sinuous**,   49

43.

Silver spots on under side of wings,   Agraulis Vanillæ, (36)
No silver spots on under side of wings,   44

44.

Three large "eye-spots" on the upper surface, Junonia Cœnia, (66)
Five or six small eye-spots, with or without pupils, on the
upper surface of the hind wings,   Apatura, 46
Upper surface without eye-spots,   45

45.

Fulvous, with a paler mesial band,   Euptoieta Claudia, (46)
Fulvous, marked with four somewhat united basal lines, and
three bands on the fore wings that are partially united,

Eresia Frisia, (54)

46.

One eye-spot near outer margin of fore wings,   48
No eye-spots on fore wings,   47

47.

Two outer rows of pale spots lighter than the ground color,

Apatura Clyton, (78)

Two outer rows of spots not lighter than the ground color

Apatura Flora, (79)

48.

Hind wings fulvous,   Apatura Alicia, (77)
Hind wings dark brown,   Apatura Celtis, (76)

49.

General color (except ♀ of *Diana*) fulvous, under side of wings
    (except *Bellona*) with silver spots,   Argynnis, 50
General color black, with a terminal border of red spots,

Melitæa Phaeton, (47)

General color fulvous, with a prominent black border; no
    silver spots on the under side,   ·   57

50.

Under side of wings with silver spots,   51
Under side of wings without silver spots, though white spots
    may be present,   56

51.

Basal two-thirds of wings of ☿ dark fulvous brown, outer
    third fulvous; ♀ blue-black, outer third blue,

Argynnis Diana, (38)

Fore wings fulvous, hind wings black,   Argynnis Idalia, (37)
Both wings fulvous,   52

52.

Basal part of wings plain fulvous brown,   53
Basal part of wings not differing from the rest, size medium,

Argynnis Myrina, (43)

53.

Basal half of wings fulvous brown,   Argynnis Cybele, (39)
Less than half of wings brown,   54

54.

Under side of hind wings with a light submarginal band,   55
Under side of hind wings without a light submarginal band,

Argynnis Alcestis, (41)

55.

Under side of hind wings light brown,   Argynnis Aphrodite, (40)
Under side of hind wings maroon,   Argynnis Atlantis, (42)

56.

With white spots on the under side,   Argynnis Montinus, (44)

Under side without white spots,   **Argynnis Bellona**, (45)

57.

Under side of hind **wings fulvous**; a central and basal band of
  buff spots; submarginal lunules white, **Mellitæa Harrisii**, (48)
Under side of hind wings brownish or brownish yellow,

            Phyciodes, 58

58.

Broad central band of white or light buff on under side, 59
Central band of under side narrow or wanting, submarginal
  row of spots small,   60

59.

Submarginal row of dark brown spots on under side of hind
  wings pupilled with white, **Phyciodes Nycteis**, (49)
Submarginal row of spots on under side with no more than one
  pupilled with white,   **Phyciodes Carlota**, (50)

60.

Upper side of fore wings with a pale fulvous or almost white
  band beyond the cell,   61
Upper side of fore wings without a whitish band,

          Phyciodes Tharos, (52)

61.

Expanse from 1 to 1.25 inches; a black terminal patch on
  under side of hind wings, **Phyciodes Phaon**, (51)
Expanse from 1.25 to 1.5 inches; no black terminal patch on
  under side of hind wings, **Phyciodes Batesii**, (53)

62.

Hind wings tailed,   63
Hind wings not tailed,   64

63.

Hind wings with two prominent tails, apex of fore wings
  truncate,   **Timetes Petreus**, (69)
Hind wings with one tail and a prominent angle, apex rounded,

        Victorina Steneles, (70)
Hind wings with one short tail, apex pointed,

        Paphia Troglodyta, (80)

64.

Gray; two round black spots on hind wings, one on fore wings,

        Anartia Jatrophæ, (67)
Color not gray,   65

65.

Ground color dark chocolate with bluish-purple reflections, **two**
  white patches on fore wings and one on hind wings,
                                    Diadema Misippus, (71)
Ground color purplish brown,   **Eunica Monima,** (68)
Ground color black or fulvous, or mahogany-brown,
                                        Limenitis, 66

66.

Ground color black,   67
Ground color fulvous or mahogany,   68

67.

Wings without broad white bands,   **Limenitis Ursula,** (72)
Both wings with broad white bands, **Limenitis Arthemis,** (73)

68.

Upper side fulvous, under side paler, **Limenitis Disippus,** (74)
Both surfaces mahogany-brown,   **Limenitis Eros,** (75)

69.

A golden or silver spot on under side of hind wings,
                                        Grapta, 70
No golden or silver spot on under side of hind wings,   75

70.

Spots in the discal cell of fore wings wholly or partly separate, 71
Spots in the cell of fore wings blended into a transverse dash,
                                        Grapta J Album, (60)

71.

Silver spot on under side of hind wing in two pieces, forming
  a semicolon,   **Grapta Interrogationis,** (55)
Silver spot single,   72

72.

Fore wings with a subterminal row of three round black spots,
  the lowest double,   73
Fore wings with a subterminal row of three round black spots,
  the lowest single,   74

73.

Silver mark a distinct comma,   **Grapta Comma,** (56)
Silver mark an open L,   **Grapta Gracilis,** (58)

e   6*

74.

Under side of wings fawn color, marked with brown and lilac,

Grapta Faunus, (57)

Under side dark brown ; a light band across the fore wings,

Grapta Progne, (59)

75.

Apex of fore wings distinctly truncate, the angles sharp,

Vanessa,   76

Apex of fore wings somewhat truncate, the angles rounded,

Pyrameis,   77

76.

Ground color maroon-brown, outer margin yellow, supple-
    mented by a row of blue spots,   Vanessa Antiopa, (61)

Ground color brown, submarginal band fulvous,

Vanessa Milbertii, (62)

77.

Ground color black, band on fore wings and outer margin of
    hind wings fulvous,   Pyrameis Atalanta, (63)

Ground color fulvous,   78

78.

Five eye-spots on under side of hind wings,

Pyrameis Cardui, (65)

Two eye-spots on under side of hind wings,

Pyrameis Huntera, (64)

79.

Wings entire, eyes hairy or naked,   80

Wings dentate, hind wings strongly angled in the middle,
    eyes hairy,   Debis Portlandia, (81)

Hind wings dentate, eyes naked,   Satyrus,   85

80.

Under side of hind wings without thick marbling of dark-
    brown abbreviated lines, alternating with gray and brown,   81

Under side with marbling of dark-brown abbreviated lines, al-
    ternating with gray and brown,   Chionobas,   86

81.

Wings blackish brown or wood-brown,   82

Wings russety, eye-spots above prominent on both wings,

Neonympha Canthus, (82)

82.

With eye-spots above,   Neonympha Eurytris, (85)
Without eye-spots above,   83

83.

With two black marks on the outer margin of hind wings
     above,   Neonympha Gemma, (83)
Without marks above,   84

84.

Transverse lines on the under side dark brown,
       Neonympha Sosybius, (86)
Transverse lines on the under side ferruginous ochre,
       Neonympha Areolatus, (84)

85.

Fore wings with a buff band containing one ocellus, or one
     ocellus and a small black spot,   Satyrus Pegala, (87)
Fore wings with or without a buff band, but with two ocelli,
       Satyrus Alope, (88)

86.

With eye-spots above,   Chionobas Jutta, (89)
Without eye-spots above,   Chionobas Semidea, (90)

87.

Brownish fulvous or brown; numerous rows of small black
     spots,   88
Black and green; abdomen orange,   Eumenia Atala, (94)

88.

Brown; expanse 1 inch or more,   Calephelis Borealis, (93)
Brownish fulvous; expanse .8 of an inch,   Calephelis Cænius, (92)

89.

Palpi projecting in front scarcely the length of the head; an-
     tennæ reaching to the middle of the fore wings; colors
     blue, black, or blackish brown,   Thecla,   90
Palpi projecting in front more than the length of the head;
     antennæ not reaching to the middle of the fore wings   107

90.

Hind wings with two slender tails, or an angle in place of the
     second,   91

Outer margin of hind wings dentate; no orange crescents beneath,   104
Hind wings entire, anal angle sharp,   106

91.

Upper side of wings blue, with a black border,   92
Upper side of wings dusky or blackish brown,   94
Upper side black, more or less tinged with blue; a red line beneath edged with white,   Thecla Poeas, (108)

92.

Under side of abdomen orange,   Thecla Halesus, (95)
Under side of abdomen not orange,   93

93.

Dark blue; border one-fourth the length of wing,
                                       Thecla M Album, (96)
Pale blue; border more than one-fourth the length of wing,
                                       Thecla Acis, (107)

94.

A pale-blue patch between two orange spots on under side of
    hind wings near anal angle,   95
No blue patch near anal angle; under side green marked with
    brown and white,   Thecla Smilacis, (106)

95.

Upper surface with orange at anal angle,   96
Upper surface without orange, or at least very faint,   98

96.

Orange on hind wings, a crescent enclosing a black spot,
                                       Thecla Humuli, (99)
Orange, a patch or band not clearly defined,   97

97.

The points of the W formed by the inner line on the under
    side of the hind wings touching the outer line,
                                       Thecla Favonius, (97)
The points of the W not touching the outer line,
                                       Thecla Autolycus, (98)

98.

Under side pale bluish gray,   Thecla Acadica, (100)
Under side not pale gray,   99

99.

Under side with three or more whitish stripes,

Thecla Strigosa, (105)

Under side with not more than two stripes, 100

100.

Color dark brown, 101

Color yellowish brown, **Thecla Edwardsii**, (101)

101.

Marks at the ends of cells on under side of both wings, 102

No marks at the ends of cells, **Thecla Ontario**, (104)

102.

Under side ash-gray, **Thecla Columella**, (109)

Under side brown, 103

103.

Inner line on the under side double, **Thecla Calanus**, (108)

Inner line on the under side single, **Thecla Wittfeldii**, (102)

104.

Brown beneath; outer half lighter, 105

Dark brown; under side with two light bands edged with white on the hind wings and one on the fore wings,

Thecla Niphon, (113)

105.

A fine dark-brown line separating the two colors of the under side, **Thecla Augustus**, (110)

A white line marking the separation; first tooth outside of anal angle curving outward, **Thecla Irus var. Arsace**, (111)

A white line marking the separation; first tooth outside of anal angle not curving outward, **Thecla Henrici**, (112)

106.

Under side of hind wings with one row of orange spots,

Thecla Titus, (115)

Under side of hind wings with two rows of orange spots,

Thecla Læta, (114)

107.

. Color above fulvous, or coppery and black, 108

Color above blue, or bluish black, **Lycæna**, 112

108.

Wings fulvous; border of fore wings and base of hind wings
black,   **Feniseca Tarquinius,** (116)
Wings coppery, or purplish **black with fulvous bands,**
                                   Chrysophanus, 109

109.

Color above black with a coppery tinge; under side gray, with
black points,   110
Color distinctly coppery; orange border on hind wings above
and below,   111

110.

Size medium; half row of orange crescents on under side of
hind wings,   **Chrysophanus Dione,** (117)
Size small; crescents faint,   **Chrysophanus Epixanthe,** (119)

111.

Size medium,   **Chrysophanus** Thoe, (118)
Size small,   **Chrysophanus** Hypophleas, (120)

112.

Hind wings without tails,   113
Hind wings with a slender tail; color blue, or black tinged with
blue; hind wings with two orange crescents,
                                 Lycæna Comyntas, (124)

113.

Upper surface blue,   114
Upper surface not blue,   118

114.

Under side with black spots, having metallic scales near the
anal angle,   115
Under side without such spots,   116

115.

Upper side blue, narrow or broad border, or almost **black; one
anal spot on** under side circled with metallic **green scales,**
                                 Lycæna Filenus, (125)
Upper side violet-blue; two **or three anal spots; under side of**
hind wings with two black spots **circled with metallic** blue,
                                 Lycæna Ammon, (128)
Upper side almost white, with broad border, or pale violet-blue;
under side white, with many dark lines; anal spot circled
with blue,   **Lycæna Theonus,** (129)

116.

Under side of hind wings with a border of metallic spots,

Lycæna Scudderii, (122)

Under side without metallic spots,   117

117.

Under side dark gray,   Lycæna Lygdamus, (121)

Under side pale bluish gray,   Lycæna Pseudargiolus, (123)

118.

Upper surface black,

Lycæna Pseudargiolus, var. Violacea, ♂ (123)

Upper side brown, under side brown, with numerous white
bands,   119

119.

Six subterminal round black spots on under side of hind wings,

Lycæna Isophthalma, (126)

Seven subterminal round black spots on under side of hind
wings,   Lycæna Exilis, (127)

120.

Knob of antennæ thick, ovoid, or elongate ovoid,   121

Knob of antennæ spindle-shaped,   185

121.

Knob of antennæ without a hook or bent projection at the end, 122

Knob of antennæ ending in a hook or bent projection,   126

122.

Knob blunt, conical, without a spine,   123

Knob elongate or ovoid, rounded at the tip, straight or a little
semilunar; somewhat compressed,   Pyrgus, 125

Knob rounded; the last joint ending in a short, slender spine;
fore wings brown washed with yellow; hind wings yellow, with brown border,   Ancyloxypha Numitor, (132)

123.

Third joint of palpi concealed by hair of second; abdomen
much longer than hind wings; brown, with fulvous yellow spots,   Carterocephala, 124

Third joint of palpi free; abdomen but little if any longer than
hind wings; brown; costal edge yellow,

Thymelicus Poweshiek, (133)

124.

Costal edge not yellow; marginal and abdominal rows of spots,
  and several near the base,   Carterocephala **Mandan**, (130)
Costal edge yellow, to near the end of cell; submarginal row of
  spots,   Carterocephala **Omaha**, (131)

125.

Black, with three more or less complete rows of transverse
  white spots,   Pyrgus **Tessellata**, (177)
Brownish black, with two rows of transverse white spots,
                                   Pyrgus **Centaureæ**, (178)

126.

Tip of antennæ suddenly bent, with a much contracted, pointed
  little hook, nearly half as long as the knob; third joint of
  palpi almost concealed,   **Pamphila**, 127
Antennæ similar; fringes light; black at the ends of the veins;
  abdomen thin, scarcely reaching the anal angle,
                                       Amblyscirtes, 182

127.

Hind wings yellow, with an outer border of dark brown, not
  more than one-third the length of the wing,   128
Hind wings brown, with a yellow band,   141
Hind wings without a yellow band,   155

128.

Border of hind wings less than one-fourth the length of wing,   129
Border of hind wings one-fourth the length of wing or more,   138

129.

Fore wings without a black sex-mark or stigma in the middle, 130
Fore wings with a black sex-mark in the middle,   131

130.

Fore wings with a brown patch beyond the cell; veins in the
  yellow part usually yellow,   **Pamphila Zabulon**, ♂ (135)
Fore wings without a patch beyond the cell; veins brown,
                                   Pamphila Delaware, ♂ (166)

131.

Border of hind wings a series of triangular spots,
                                   Pamphila Phylæus, ♂ (143)
Border of hind wings continuous,   132

132.

Yellow of hind wings clear; fringes fuscous,

Pamphila Sassacus, ♂ (136)

Yellow of hind wings brownish; fringes white,

Pamphila **Uncas,** ♂ (138)

133.

Fore wings with a black stigma or sex-mark in the middle, 134

Fore wings without a sex-mark, 139

134.

Fore wings with a brown patch connected with the upper end
of stigma, 135

Fore wings with a brown oblique ray connected with the upper
end of stigma, 138

Fore wings with the brown subapical patch not connected with
the stigma; stigma and spot containing it nearly circular,

Pamphila Huron, ♂ (142)

135.

Under side ferruginous brown, 136

Under side not ferruginous brown, 137

136.

Yellowish-white bands on under side of both wings,

Pamphila Leonardus, ♂ (140)

Yellowish-white bands on under side of fore wings,

Pamphila Meskei, ♂ (141)

137.

Under side dark brown, overlaid with yellow scales and a yellow band, Pamphila Brettus, ♂ (144)

Under side yellow, with a paler yellow band,

Pamphila Mystic, ♂ (147)

138.

Under side of hind wings dark brown sprinkled with dark yellow scales, with yellow band but no yellow ray before
inner margin, Pamphila Pontiac, ♂ (161)

Under side of hind wings dark yellow, a paler ray from cell
out, and one before inner margin, Pamphila Dion, ♂ (162)

Under side of hind wings dark brown, heavily overlaid with
russety brown; a pale ray through cell,

Pamphila Palatka, ♂ (164)

D 7

139.

Fore wings brown, a broad yellow band, and the basal part
heavily washed with yellow; under side of hind wings
fulvous yellow, with pale, obscure band,

                                    Pamphila Byssus, ♂ (167)

Fore wings with basal two-thirds yellow; under side of hind
wings brown, heavily overlaid with russety brown, with a
pale ray through the cell,   Pamphila Palatka, ♀ (164)

Fore wings yellow; under side of hind wings yellow, without
marks,   140

Fore wings dark brown, with a row of yellow spots, and a spot
in the cell,   Pamphila Viator, ♂ and ♀ (172)

140.

Fore wings yellow at the base,   Pamphila Vitellius, ♂ (165)

Fore wings dark brown at base; a bar of brown at end of cell,
                                    Pamphila Delaware, ♀ (166)

141.

Fore wings with an oblique black sex-mark,   142

Fore wings without a sex-mark, but with a row of spots beyond the middle,   143

142.

Under side of hind wings ferruginous brown, with two yellow
bands,   Pamphila Peckius, ♂ (146)

Under side of hind wings yellowish brown, with two white
bands, the basal one broken, the outer one sending a ray
outward from its posterior end,   Pamphila Metea, ♂ (137)

Under side of hind wings ferruginous brown, with one yellow
band and a spot,   Pamphila Leonardus, ♂ (140)

Under side of hind wings brown, sprinkled with yellow; five
white spots,   Pamphila Seminole, ♂ (139)

143.

Basal half of costal margin of fore wings yellow, or heavily
washed with yellow,   144

Costa not yellow, only sprinkled at least,   148

144.

The cell of fore wings yellow, except at base,   145

Cell of fore wings dark brown,   146

145.

Band on upper side of hind wings reduced to a yellow central
   patch,   **Pamphila Vitellius,** ♀ (165)
Band on upper side of hind wings not abbreviated,
                            Pamphila Mystic, ♀ (147)

146.

Under side of hind wings ferruginous brown,   147
Under side of hind wings ochre-yellow, with a paler band and
   brown clouds,   **Pamphila Phylæus,** ♀ (143)

147.

Under side of hind wings with one yellow band and a spot,
                       Pamphila Leonardus, ♀ (140)
Under side of hind wings with very indistinct band, or none,
                       Pamphila Meskei, ♀ (141)

148.

Yellow band of hind wings faint,   149
Yellow band of hind wings distinct,   .   150

149.

Under side of hind wings dark brown, with rusty brown scales
   and a continuous pale band,
          Pamphila Zabulon form Pocahontas, ♀ (135)
Under side of hind wings dark brown, washed with ochre scales
   and with two or three small spots; upper side of hind
   wings sprinkled with yellow,   **Pamphila Sassacus,** ♀ (136)
Under side of hind wings brown, washed with ochre, a subterminal white band sending a ray outward from its posterior end,   **Pamphila Metea,** ♀ (137)

150.

Under side of hind wings with two light bands,   151
Under side of hind wings with one light band or none,   152

151.

Under side of hind wings dark brown washed with pale yellow; bands white, separate,   **Pamphila Uncas,** ♀ (138)
Under side of hind wings ferruginous brown, with the two
   yellow bands united in the middle,
                    Pamphila Peckius, ♀ (146)

152.

Under side of hind wings dark brown, sprinkled with pale yellow and ferruginous scales,   153

Under side of hind wings dark brown, sprinkled with ferruginous brown,   154

Under side of hind wings dark rusty yellow; two yellow ray
stripes,   **Pamphila Dion**, ♀ (162)

Under side of hind wings fulvous yellow, with obscure pale
band,   **Pamphila Byssus**, ♀ (167)

153.

Under side of hind wings with a white band and two or three
unconnected spots,   **Pamphila Huron**, ♀ (142)

Under side of hind wings with a band of five small white spots,
and a bar across cell of fore wings above,
                    Pamphila Seminole, ♀ (139)

154.

Under side of hind wings with a band of five elongated yellow
spots,   **Pamphila Pontiac**, ♀ (161)

Hind wings with three whitish spots above and below,
              Pamphila Ethlius, ♂ and ♀ (159)

155.

Fore wings with an oblique stigma or sex-mark,   156

Fore wings without a sex-mark,   166

156.

Fore wings with the basal half more or less yellow,   157

Fore wings without distinct yellow on the basal half,   161

157.

Washed with greenish yellow, a large subquadrate black patch
below stigma sending a spur towards posterior angle,
           Pamphila Otho var. Egeremet, ♂ (145)

Patch below stigma only moderate,   158

158.

Cell of fore wings and costa beyond cell clear yellow,   159

Costal margin not clear yellow,   160

159.

Under side of hind wings sprinkled with yellow, but without
bands,   **Pamphila Cernes**, ♂ (148)

Under side of hind wings sprinkled with pale yellow, with an
obscure discal whitish band,   **Pamphila Myus**, ♂ (149)

160.

Costal margin of fore wings with six or seven yellow rays
above and beyond the cell; under side of hind wings yellow; no spots, **Pamphila Arpa,** ☿ (163)

Costal margin somewhat washed with yellow, but without rays;
under side of hind wings brown, sprinkled with yellow;
no spots, **Pamphila Bimacula,** ☿ (160)

161.

Fore wings with a row of four or more white spots beyond the
middle, 163

Fore wings without white spots, olivaceous brown, 162

162.

With a row of faint spots on under side of both wings,
Pamphila Metacomet, ☿ (153)

Without a row of faint spots on under side of wings,
Pamphila Vestris, ☿ (152)

163.

Spots in three subcostal interspaces and one in first median interspace, 164

Three spots in subcostal and two in median interspaces, and one
in submedian, 165

164.

Outer part and basal portion of under side of hind wings lilac,
with brown cloudings and a faint purplish band,
Pamphila Accius, ☿ (154)

Outer part of under side of hind wings gray, with no bands,
Pamphila Hianna, ☿ (171)

165.

Fore wings with a small whitish spot in lower side of cell near
the end, **Pamphila Verna,** ☿ (151)

Fore wings with no spot in the end of cell,
Pamphila Manataaqua, ☿ (150)

166.

Fore wings with a yellow or whitish spot in the end of cell, 167

Fore wings with two spots in the end of cell,
Pamphila Loammi, ♀ (155)

Fore wings without a spot in the cell, 171

167.

Spot in anterior part of cell near the subcostal vein,   168
Spot in posterior part of cell near the median vein,
                    Pamphila Panoquin, ♂ and ♀ (157)

168.

Spot in cell white, rounded,   169
Spot a yellow ray, six yellow spots in the discal row,   170

169.

Under side of hind wings gray, a ray near inner margin without gray; a very faint whitish band,
                    Pamphila Eufala, ♂ and ♀ (169)
Outer and basal part of under side of hind wings lilac, anterior part brown clouded; an indistinct purple band,
                    Pamphila Accius, ♀ (164)
Outer part of under side of hind wings gray, with no bands,
                    Pamphila Hianna, ♀ (171)

170.

Under side of hind wings sprinkled with yellow, without a
    band,   Pamphila Cernes, ♀ (148)
Under side of hind wings sprinkled with pale yellow, with dim
    whitish discal band,   Pamphila Myus, ♀ (149)

171.

Wings without marks,   Pamphila Fusca, ♂ and ♀ (170)
Fore wings with a more or less distinct outer row of spots
    extending to submedian vein,   172
Fore wings with more or less of an outer row of spots, but not
    reaching submedian vein,   176

172.

A band on under side of hind wings, but not very distinct,   173
No band on under side of hind wings,   175

173.

Outer row of fore wings with two spots beyond the cell; under
    side of hind wings lilac along outer margin,
                    Pamphila Zabulon form Quadraquina, ♀ (135)
No spots beyond the cell of fore wings,   174

174.

Under side of hind wings rusty brown,

Pamphila Otho, ♂ and ♀ (145)

Under side of hind wings dark reddish brown, sprinkled with
pale yellow scales, Pamphila Manataaqua, ♀ (150)

175.

Under side of hind wings yellow, Pamphila Arpa, ♀ (163)

Under side of hind wings dark brown, the veins gray,

Pamphila Ocola, ♂ and ♀ (158)

176.

Under side of hind wings without spots or bands, spots above
not very distinct, 177

Under side of hind wings with a more or less distinct row of
spots, 78

177.

Under side of hind wings dark brown sprinkled with gray,

Pamphila Osyka, ♂ and ♀ (168)

Under side of hind wings reddish brown, Pamphila Vestris, ♀ (152)

Under side of hind wings dark brown sprinkled with dusky
yellow, Pamphila Bimacula, ♀ (160)

178.

Under side of hind wings with three white spots, two of which
are contiguous, Pamphila Maculata, ♂ and ♀ (156)

Under side of hind wings with a faint row of pale spots, 179

Under side of hind wings with distinct yellow band, 181

179.

Under side of body and antennæ distinct greenish,

Pamphila Otho var. Egeremet, ♀ (145)

Under side of body gray, 180

180.

Fore wings with an outer row of five distinct white spots, the
fifth twice as large as the fourth, Pamphila Verna, ♀ (151)

Fore wings lacking the anteapical spots of the outer row, or the
merest trace of one spot, if any; band on under side of hind
wings distinct, Pamphila Metacomet, ♀ (153)

181.

Under side of hind wings yellow ferruginous; a yellow band,
one stripe of which extends as a ray to middle of cell,

Pamphila Massasoit, ♂ and ♀ (134)

Under side of hind wings dull olivaceous, with clouds of blackish brown; a yellow band towards outer margin,

Pamphila Brettus, ♀ (144)

182.

Fore wings with no distinct spots except the anteapical,   183
Fore wings with more than the three anteapical spots,   184

183.

Under side of hind wings washed with lilac, with no clearly
defined row of spots, but with a row of whiter clouds;
fringe alternate gray and fuscous, Amblyscirtes Vialis, (173)
Under side of hind wings washed with chalky scales, with a
row of whitish points, one in the cell and two above the
cell; fringes white and fuscous,   Amblyscirtes Eos, (174)

184.

Under side of hind wings with an outer row of connected pale
yellow spots, and one in the cell, Amblyscirtes Samoset, (175)
Under side of hind wings with an outer row of spots, and one
below the cell,   Amblyscirtes Textor, (176)

185.

Palpi gray or brownish below,   ·   186
Palpi white below,   203

186.

Palpi surpassing the front by more than the length of the eyes,
Nisoniades, 187
Palpi surpassing the front by less than the length of the eyes, 196

187.

Fore wings without a subterminal row of white transparent
spots,   188
Fore wings with a subterminal row of white semi-transparent
spots; four near the costa and the others in median interspaces,   191

188.

Upper side of fore wings not overlaid with pale bluish scales
on the outer part,   189
Upper side of fore wings heavily overlaid with pale bluish
scales on the outer part,   190

189.

The discal cross-vein of fore wings not conspicuously marked in
brown,   Nisoniades Brizo, (179)

The discal cross-vein of fore wings conspicuously marked in
brown,   - **Nisoniades Ausonius**, (184)

190.

Outer row of large spots overlaid with light in the centre so as
to make them appear as an irregular row of black rings,
Nisoniades Icelus, (180)

Outer row of large spots only slightly sprinkled with pale
scales,   **Nisoniades Somnus**, (181)

191.

Upper surface umber-brown,   192
Upper surface dark brown,   194

192.

Dark spot in base of cell of fore wings and the two rows of
spots very distinct,   **Nisoniades Martialis**, (185)

The spot in base of cell of fore wings not distinct,   193

193.

The mesial band distinct,   **Nisoniades Lucilius**, (182)
The mesial band not distinct,   **Nisoniades Persius**, (183)

194.

No pale umber patch at the end of cell of fore wings,
Nisoniades Juvenalis, (186)

A pale umber patch at the end of cell of fore wings,   195

195.

Umber patch at the end of cell only,   **Nisoniades Nævius**, (188)

Umber patch at the end of cell, between median and submedian
veins and near posterior angle,   **Nisoniades Petronius**, (187)

196.

Under side of hind wings not banded with brown and purple,
Eudamus, 197

Under side of hind wings banded with brown and purple,
Erycides, 206

197.

Hind wings rounded at the anal angle,   198
Hind wings more or less produced or tailed at the anal angle, 201

198.

Fore wings with white spots,   199
Fore wings with yellow bands,   200

f

199.

White spots mere points on the costa and in the median interspaces,   Eudamus Pylades, (191)
White spots larger, forming almost continuous bands,
                                    Eudamus Bathyllus, (192)

200.

Outer part of under side of hind wings white,
                                    Eudamus Lycidas, (193)
Outer part of **under side** of hind wings not white,
                                    Eudamus Cellus, (194)

201.

Hind wings produced, not tailed,   202
Hind wings tailed,   Eudamus Proteus, (197)

202.

Under side of hind wings with a white band in the middle,
                                    Eudamus Tityrus, (196)
Under side of hind wings without a white band in the middle,
                                    Eudamus Zestos, (195)

203.

Palpi surpassing the front by about the length of the eyes;
    small species,   Pholisora, 204
Palpi surpassing the front by less than the length of the eyes;
    large species,   205

204.

Outer margin entire,   Pholisora Catullus, (189)
Outer margin dentate,   Pholisora Hayhurstii, (190)

205.

Without yellow markings above,   Erycides, 206
With yellow markings above,   Megathymus, 207

206.

Upper side without white spots,   Erycides Batabano, (198)
Upper side with white anteapical spots, also at end of cell and
    in median interspaces,   Erycides Amyntas, (199)

·   207.

Hind wings with a yellow border only on the anterior part of
    the outer margin, if present at all, Megathymus Yuccæ, (200)
Hind wings with a yellow border reaching to the submedian
    vein at least,   Megathymus Cofaqui, (201)

FAMILY PAPILIONIDÆ.

THE butterflies of this family, the highest of the Lepidoptera, may be known by their broad wings, which are
erect in repose; the antennæ are slender, the knob either
straight or slightly curved; the body slender and furnished with six feet fitted for walking, the first pair of
normal size and directed forward, the other two pairs
directed outward and backward. The larvæ are smooth,
or only moderately provided with short hairs or pile,
never provided with branching or simple spines, and in
only one instance—*Papilio Philenor*—provided with
fleshy protuberances. The chrysalides are naked, the tip,
or cremaster, fastened to a button of silk, and the body
suspended obliquely in a loop of silk that passes round
it a little in front of the middle. The family is divided
into two subfamilies, Papilioninæ and Pierinæ.

SUBFAMILY PAPILIONINÆ.

This contains only the genus Papilio. These are
large butterflies, often known as swallow-tails. The
wings are ample; the fore wings triangular; the hind
wings concave next to the body, and usually provided
with a tail-like appendage before the anal angle; the
outer margin dentate, with the teeth quite prominent
near the tail. The larvæ are smooth, or with a few
short scattered hairs; spindle-shaped, being thickest
through about the fourth segment, *P. Philenor* being
provided with four rows of slender fleshy processes.
In the upper anterior part of the second segment is a

forked scent-organ which is capable of being thrust out
at will or withdrawn into a slit-like receptacle. When
extended, this organ gives off a disagreeable odor which
serves as a protection to the larva.

1. PAPILIO AJAX, Linn.

Expanse of wings from 2.6 to 3.5 inches.

Upper surface of body and wings varying from pale
black to deep black, banded and marked with a color varying from greenish yellow to pale blue-green. These are
arranged in three bands common to both wings, the one
next to the body narrow, the third bifid on the fore
wings from the costa to the median vein; and three
short bands not reaching farther than the median vein.
There is also a subterminal band on the fore wings
crossed by the black veins, and a row of spots on the
hind wings parallel with the outer margin. Body
blackish, with two whitish lines on the sides.

There are three dimorphic forms of this species, and
one variety. The first of these is the winter form,
WALSHII, Edw.—In this the black is pale, the light
parts greenish yellow; the tail to the hind wings about
.6 of an inch long, black, tipped with light; and the
red before the anal ocellus is a bent bar; after the ocellus
are two blue lunules.

Var. ABBOTTII, Edw.—Expanse from 2.6 to 2.8
inches. This resembles the above, but has a more or
less distinct crimson streak on the hind wings nearly
parallel with the internal margin.

Winter form, TELAMONIDES, Feld.—Expanse from
2.8 to 3.2 inches. This is about the same in color and
markings as form *Walshii*, but the tail is a little longer,

and the outer end is not merely tipped with greenish yellow, but bordered on each side from half to two-thirds the distance from the tip to the base, and the anteanal crimson bar is sometimes two spots.

Summer form, MARCELLUS, Bd.—Expanse from 3.2 to 3.5 inches. This is black, with the light part blue-green; the tail over an inch long and bordered with yellow, and the anteanal crimson mark one or two spots instead of a bar.

All of these forms have the markings of the upper side repeated beneath, with a more or less prominent crimson stripe through the middle of the hind wings nearly parallel to the inner margin.

Of these three forms the last is the one found in summer, and comes from chrysalides formed the same season, while the others emerge from chrysalides that have wintered over. They were originally described as separate species, and were generally considered so till Mr. Wm. H. Edwards proved their identity by breeding the different forms.

The eggs are deposited on the leaves of the pawpaw, upon which the larvæ feed. They are pale green, globular, smooth, .016 of an inch in diameter. The young larva is black, covered with minute papillæ, from each of which proceed fine hairs. After the first moult it is ash-colored, still covered with the papillæ. These are lost at the second moult, when the larva assumes the general form and smooth skin which it shows at maturity; the color yellowish white, with transverse gray stripes. After the third moult the color is smoky brown, each segment crossed by four lines, of which the anterior is yellow and the rest white; the second, third, and fourth segments without **yellow; at the** junction of the fourth and fifth joints is **a** velvety-black band, preceded by white and followed **by** yellow. **At** the fourth moult the color becomes **darker, each** segment crossed **by** a pale yellow and four gray stripes.

In some of the larvæ the general color is gray, with white, black, and yellow bands on the fourth and fifth segments, and the segments after the fifth crossed by one yellow **and** two dull white stripes. In **other specimens** the color is blue-green, each segment crossed **by gray,** yellow, and clear white, the white band replaced by turquoise-blue. In still others the color is pale green throughout, except one yellow stripe on each segment, the bands being blue, black, and yellow.

Chrysalis from .8 to .9 of an inch long, cylindrical, tapering posteriorly from the middle of the abdomen; head triangular, terminated by two short lateral points; another, beak-like, **on the thorax; from** this two small ridges pass along the wing-cases **and** down the abdomen to the extremity, and between **them two** others starting from the upper segments **of the** abdomen, on the outer sides of which last, in those chrysalides which **are brown,** is a fine light-colored line. Color dead-leaf brown or bright green.

Location from Pennsylvania to Texas, **and** through the Mississippi Valley, where it flies from March or April till cool weather in the fall.

2. PAPILIO PHILENOR, Linn.

Expanse of wings **from 3.5 to 4.5 inches.**

Body and wings black; the hind wings of the male reflecting a metallic green, those of the female a steel-blue.

Parallel with the outer margin of the fore wings there is a row of more or less distinct spots running from the posterior angle half-way to the apex ; on the hind wings are six whitish spots between the venules, the one before the anal ocellus very small. Tail about .3 of an inch.

On the under side the marginal spots on the fore wings are more distinct, as are also the yellowish crescents in

FIG. 11.

Papilio Philenor (natural size).

the fringe. The metallic sheen of the hind wings occupies the outer half of the wing, and contains a crescent of seven orange spots, each bordered with white on its costal side, and more or less completely with black the rest of the way.

The egg is spherical, the surface much covered with a rough crust, which rises to a summit, either small and pointed, or rather large and truncated; the sides of this crust irregularly melon-ribbed. Color of the surface russet, of the crust bright ferruginous.

The young larva is ferruginous, marked longitudinally by many rows of low, conical, black tubercles, each supporting a black hair. When mature (Fig. 12), it is two inches long; color velvety black, with a slight purplish or chestnut-brown hue; covered with long fleshy tubercles of the same color as the body, and shorter orange-colored tubercles, as follows: two brown ones on joint 2; two brown ones and two orange ones on joint 3; joints 4 and 6 the same; joint 5 with four orange tubercles; joints 7 to 10 each with two brown lateral tubercles and two orange ones; joints 11 and 12 with four brown tubercles that often have orange bases; joint 13 with two dorsal brown tubercles but no lateral ones.

Fig. 12.

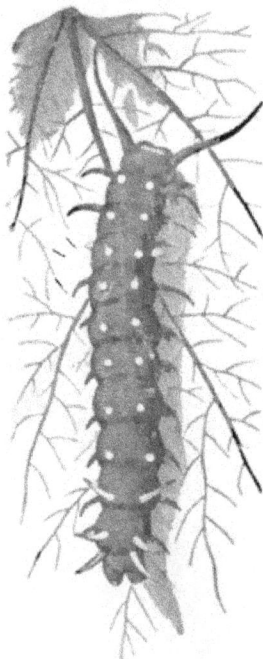

P. Philenor, larva.

Joints 8 to 11 have each a lateral orange spot just before and above the spiracles, which are sunk into the flesh and scarcely perceptible. Head, legs, and venter the same color as the body; the top of joint 2 with an orange transverse spot on the anterior edge, from which is thrust out the osmaterium, or scent-organ.

The chrysalis, represented in Fig. 13,—a, shaded back view; b, lateral outline,—is at first yellowish green, but soon becomes marked with gray and violet, with more or less yellow on the back.

This insect feeds on the different species of Aristo-

FIG. 13.

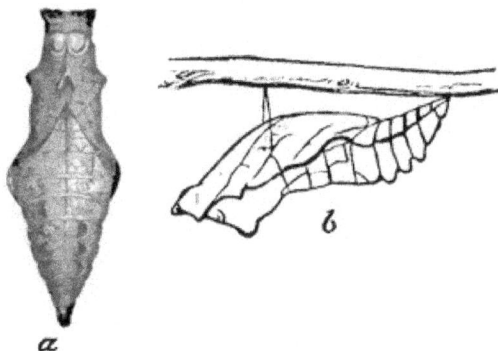

P. Philenor, chrysalis.

lochia, or pipe-vine, and is usually abundant where these plants are found. The larvæ are to be found in groups on the leaves in July and August; the pupal period lasting about three weeks.

3. PAPILIO ASTERIAS, Fab.

Expanse of wings from 3 to 4 inches.

Upper surface of body and wings black. The fore wings have two rows of yellow spots parallel with the outer margin, eight spots in each row, more prominent in the male than in the female. There are one or two spots before the inner row towards the apex, and in the male a bar at the end of the discal cell. The fringe is black, cut with yellow opposite the spots of the two rows. The spots are continued across the hind wings, the outer row being lunate, with a more or less complete row of blue clouds between the rows of yellow spots. At the anal angle there is an orange ocellus with a posterior or outer border of yellow, and a central black spot. Fringes as on the fore wings. Tail black, about .3 of an inch long.

On the under side the spots are repeated, those on the hind wings washed with orange. The body has a row

Fig. 14.

Papilio Asterias, male.

of yellow spots on each side, which continue as partial rings on the under side of the abdomen.

Var. ASTEROIDES, Reak.—This form is marked nearly as in the typical *Asterias*. In the male the inner row of yellow spots of the fore wings is almost obsolete, except the spot on the hind margin, which is prolonged into a dash. Hind wings as in *Asterias* female, but the blue clouds are reduced to small rounded patches; tails not so long as in the typical form. Below, a discal row of large fulvous sagittiform spots on the fore wings. Hind wings as in the typical form.

Aberr. Calverleyii, Grote.—Two specimens of this singular form have been taken, one a male, in August, 1863, by Mr. Louis Fischer, in the neighborhood of New Lots, Queens County, Long Island, and another, a female, in April, 1869, by Mr. T. L. Mead, near Enterprise, Florida; both being suffused forms, probably caused by the action of cold on the chrysalides soon after pupating.

In the male the upper surface has the basal two-thirds black without marks, and the remainder of the wings yellow, a narrow outer margin, and tail black. The boundary between the black and yellow on the fore wings is dentate, with the black extending out on the veins. The hind wings have a narrow subterminal crenate orange line, and an orange patch in place of the ocellus.

The under side is like the upper, except that on the hind wings there are elongate orange patches between the veins, leaving narrow yellow lines along the veins and between the ends of these patches and the black on the basal part, the subterminal line being dusky. The orange extends a little on to the fore wings as a partial terminal border.

The female is like the male, except that there is more black along the veins; the outer margin of the wing is more broadly bordered with black; there is a yellow bar at the end of the cell of the fore wings; the hind wings have two rows of orange intervenular patches in the yellow field, the inner round and the outer elongate triangular, with gray spaces between; and the ocellus has a few black scales.

The under side is like the upper, except that there are

two discal bars, with yellow on the subcostal vein, and the orange patches of the hind wings are larger.

The eggs are of a delicate light yellow, smooth and round, with the exception of being slightly flattened where they are attached to the leaf. These are deposited on the leaves of parsley, celery, parsnips, and other related plants upon which the larvæ feed.

The young larvæ are nearly black, with a broad white band across the middle, and another on the hind part of the body; thickly beset with bristles which arise from little tubercles. The second stage differs little from the first; as also the third, though there are bright spots on the body. The fourth stage is a bright green ground color with black bands, which are broad on the middle of the segments. These bands are interrupted by brick-red spots, which are arranged in three rows on each side. The tubercles are scarcely perceptible.

The fifth stage is the mature larva. When full grown

Fig. 15.

Fig. 16

Papilio Asterias, mature larva.

Papilio Asterias, pupa.

the caterpillar is about an inch and a half long, of a bright green color, with a transverse black band on each segment containing a row of yellow spots. The scent-organ in this species is yellow.

The chrysalis is an inch and a quarter long, of a pale green, ochre-yellow, or ash-gray color, with two short ear-like projections above the head, just below which, on the upper part of the back, is a little prominence. This chrysalis, like all the Papilios, is attached at the tip by a button of silk, and supported by a loop round the middle of the body. The last brood winter in the chrysalis state.

United States generally.

4. PAPILIO TROILUS, Linn.

Expanse of wings from 3.5 to 3.75 inches.

Upper surface of body and wings black, the fore wings crossed by a marginal row of greenish-yellow spots ; in some specimens a part of a second row extends from the hind margin forward. No spot in cell. Hind wings with the marginal lunules and an inner row of spots forming a broad macular band, all blue in the female and blue-green in the male. The costal spot of this inner row is mostly orange. There is a row of blue clouds between these two rows. Ocellus part orange, not pupilled. Tail .4 of an inch long.

On the under side the spots are more prominent : nearly two full rows of blue-green spots on the fore wings, and two rows of orange spots on the hind wings. Body black, with two rows of yellow spots on the sides.

The larva when full grown is a little more than one and a half inches long, the body thickest from the third to the fifth segment. It is bright green above, a yellow stripe edged behind with black across the anterior part of the second segment. On joint 4 are two prominent yellow ocelli annulate with black, and a large pupil filling most of the lower portion ; a line of black in front of the segment, and a pale pinkish spot above, margined with darker. On the fifth segment are two more ocelli. Segments 6 to 11 have each four blue dots margined with black, and there is a yellow line along the sides of the body edged with black below.

The chrysalis is 1.3 inches long, shaped in general as the allied species, the two prominences on the head projecting forward and outward, and from each of these extends a ridge along the side to the anal extremity containing a slight projection opposite the dorsal pronotal elevation. Color above the ridge pinkish orange clouded and spotted with blackish brown, there being a dorsal line, and on the abdomen four round spots to each joint. Below the color is mostly brown, veined and clouded with yellowish.

This species feeds on the leaves of spice-bush and sassafras.

Atlantic, Southern, and Western States.

5. PAPILIO PALAMEDES, Drury.

Expanse of wings from 3.5 to 4 inches.

Upper surface olive-black; the fore wings crossed by two rows of prominent yellow spots, the costal three of the inner row nearer the margin than the others, and having another spot standing before them; a bar at the end of the discal cell. Hind wings with an outer row of yellow lunules and a band corresponding to the inner row of the fore wings. Between these there is a more or less complete row of blue clouds, this space somewhat washed with yellow: the more yellow the less blue. Anal ocellus orange, partly bordered with yellow, not pupilled; some orange in the band near the internal margin. Tail .4

of an inch long, black, with a central ray of yellow. Body black, with a yellow lateral stripe.

On the under side the fore wings are about as above; the hind wings have the band white, with orange clouds on its outer edge between the veins, and each lunule has a broad dash of orange; the blue clouds more prominent than above. There is also a dull yellow stripe across the wings nearly parallel with the inner margin.

The egg is spherical, a little flattened at the base. The color is greenish yellow. It hatches in five days.

The young larva is .1 of an inch long; cylindrical, greatly thickened from joints 3 to 6, from 6 tapering to 12, then thickening to the end. There are eight rows of fleshy processes, those at the ends being larger than the others. Color of body brownish yellow marked with white; a white band, not very clearly defined, passes along the sides of segments 3 to 8; segments 12 and 13 white. It moults in four days.

After the first moult, its length is .33 of an inch. In this stage the two subdorsal rows of tubercles, or fleshy processes, are minute on joints 6 to 10; the whole of the two dorsal rows minute. Color yellow-brown, darkest posteriorly; white marks as before.

In two days it moults again, when it is .36 of an inch long; the same general shape as before. Joint 3 is a little excavated on the anterior part of the dorsum; on the second is a square-topped ridge, but the processes have disappeared. On joints 12 and 13 the processes are as during preceding stage, but the rest of the dorsal and subdorsal have disappeared. Color yellow-brown to dark brown; the sides of posterior segments of a black hue; white stripes as before. During preceding stage joint 4 had a large suboval black ocellus in a narrow yellow ring. Now the front part of ocellus is velvety black, but back of this it is vitreous black.

After three days it moults the third time. The length is .8 of an inch, color about the same, the anterior segments a little darker, their surfaces finely and thickly but indistinctly dotted green; the middle segments lighter-colored and distinctly dotted green; the side-bands salmon color; the last segments a redder salmon; 13 white above base at extremity; along base of body, with a little above spiracles, a white macular band; on dorsum of 13 are two small conical white processes; on dorsum of 5 are two abbreviated bars of red lilac, one on each side in the subdorsal row, and on 6 to 10 is a small rounded lilac spot on each in same row; on side of 8 to 10 one similar spot to each; below the basal ridge is a small indistinct blue-lilac spot on each segment from 6 to 11; ocellus as before; the buff ring now open on anterior side; head greenish yellow.

In four days more it moults the last time, taking nine days from this to reach maturity. The mature larva is 1.6 inches long; cylindrical, shaped as during the preceding stages. Color dull velvety green on joints 3, 4, 5, and on 12, 13, nearly solid, but a little specked with lighter green; the other segments light and dark green in fine markings; the basal ridge whitish green; under this is a fine black line from 3 to 12, and on 6 to 11 is a subtriangular blue spot in black edging on each segment just below the line; 2 has a narrow yellow ridge in front, nearly flat on top, the curves rounded; on anterior side of this and next it is a black subdorsal dash on each side; behind the ridge is a black rough band;

the scent-organs light yellow-brown; the ocellus on the side of 4 with a vitreous black process, the circlet orange-red, having a black stripe within its anterior edge, and a blue spot on its upper side; the blue spots along the body are set in fine black rings; on the dorsum of 5 at posterior edge is a buff spot just outside the lilac spot and touching it; head olive-green.

The chrysalis is 1.4 inches long, the ventral side highly arched, the dorsum much incurved; the former narrow at summit, rounded, sides sloping. Color variable; one phase shows the whole dorsal side a delicate green, with a darker green dorsal stripe from mesonotum back; below mesonotum a subdorsal low red tubercle on each side; on either side of the abdominal segments two rows of dull lilac points; whole ventral side one shade of green, a little darker than dorsum and less yellow; lateral ridge cream color, more or less marked by a red line, which broadens on the process of head; on ventral side below head two red dots near the middle line; a series of white dots along the margins of the wing-cases; below the ridge, on last segments, are traces of blue spots.

The natural food-plant seems to be red bay, or *Persea Carolinensis*, though they readily eat sassafras.

Gulf States, Florida to Virginia.

6. PAPILIO TURNUS, Linn.

Expanse of wings from 3.5 to 4.5 inches.

Upper surface of wings clear pale yellow, costa and outer border of fore wings, and outer and posterior border of hind wings, black; the outer portion of the black along the costa suffused with yellow; the outer border

having a row of eight yellow spots on the fore wings,
and five lunules on the hind wings, the first more or less
orange; the anal ocellus orange, with yellow on the posterior part, not pupilled. The fore wings have four
black bands or stripes; the first, about one-fourth the
distance from the base to the outer margin, is continued

Fig. 17.

Papilio Turnus (natural size).

two-thirds across the hind wings, where it turns abruptly
to meet a black edging that extends along the base of
the fore wings and along the inner margin of the hind
wings to this point. The second extends from the
costa to the median vein, or sometimes beyond; the third
extends from the costa across the end of the discal cell;
the fourth, from the costa to the fifth subcostal venule or
beyond. The broad black terminal border of the hind

wings contains a series of more or less prominent inter-venular blue clouds, sometimes small and not reaching the costal end of the border, at other times suffusing most of the black, and it may be some of the yellow. Tail .5 of an inch long, black, edged on the inside with yellow.

Under side similar to the upper, but the black termi-nal borders suffused with yellow, and the lunules washed with orange, there being a little of this on the posterior part of the yellow ground color.

Body black, with a broad yellow stripe on each side.

Sometimes the ground color instead of being pale yellow is more or less tinged with dark yellow border-ing on orange, and this may be suffused with black. These are transition stages between the typical form and the black female.

Aberr. form ♀ GLAUCA, Linn.—This is black instead of yellow. In this case the spots and lunules of the outer border remain the same, but the blue clouds of the hind wings extend in a crescent band from the costa to the internal margin, preceded by a wavy black line, and more or less of the wing inside this line washed with blue. The black ground color is usually dull, so that the transverse black bands can be traced, at least on the under side.

The eggs are nearly globular, smooth; dark green when first laid, but soon change to greenish yellow, speckled with reddish brown.

The young larva is of a brownish color mottled with black, and has a large whitish spot on the middle of the back. On each side of the dorsum on the second and last three segments a tubercle, and two on each side of third and fourth. Duration of this period four days.

During the second stage, or after the first moult, the color is blackish brown, mottled with light brown or dark green, and dorsally dotted with white. Dorsal patch yellowish, tubercles black. In five days it moults the second time, when the length is .7 of an inch. The color is mottled light and dark green on the anterior and last segments; the large patch salmon-colored, as is often more or less of the last segment; tubercles and lilac spots as before; on fourth segment a round ochraceous patch appears, on which is a black ring with a lilac centre; head brown.

Moults the third time in four days, when the length is one inch; anterior segments much thickened. Color green, the salmon patch nearly and sometimes wholly lost; the spots on joint 4 pale green, central points purple; on the same segment are two small purple spots between the others; on joint 5 is a row of four purple spots, and on 9 to 11 there is one spot on each side of each. Moults

FIG. 18.

Papilio Turnus, full-grown larva.

the last time in five days. The mature larva is about 1.5 inches long, of a deep green color, paler beneath, the head reddish brown. The anterior edge of segment 2 and the posterior part of segment 5 are yellow; the anterior part of 6 is velvety black. Some examples are dark reddish brown, or blackish, with the same markings. Head above pinkish brown. In about seven days it changes to a chrysalis. This is 1.4 inches long; cylindrical, thickest at the fifth and sixth segments, and tapering rapidly to the last; shaped as in Fig. 19. Color variable. Some examples light or wood-brown striped with dark brown; others very dark, either brown or blackish; some with a few broken stripes of green.

FIG. 19.

Papilio Turnus, pupa.

The larvæ feed on a great variety of trees,—apple, quince, thorn, plum, cherry, birch, basswood, ash, alder, oak, sassafras, catalpa, willow, and tuliptree being given. The eggs are deposited singly on the leaves, and hatch in a little less than two weeks. The mature larvæ rest on the upper side of the leaf, covering it with silk and curving it up so as partially to enclose itself.

Atlantic States; Mississippi Valley to Texas.

7. PAPILIO CRESPHONTES, Cram.

Expanse of wings from 4 to 5.25 inches.

Wings above olive-black, crossed by two rows of prominent yellow spots. One row begins at the apex of the fore wings and extends across the hind wings near the base, the part on the hind wings being a band reaching from costa to inner margin. The second row begins on the costa above the end of the cell, extends outward till it meets the first row, the third spots of each row coalescing; opposite the sixth spot of the first row it is renewed, and extends in three spots to the posterior angle. From the apex of the hind wings it is continued to the

FIG. 20.

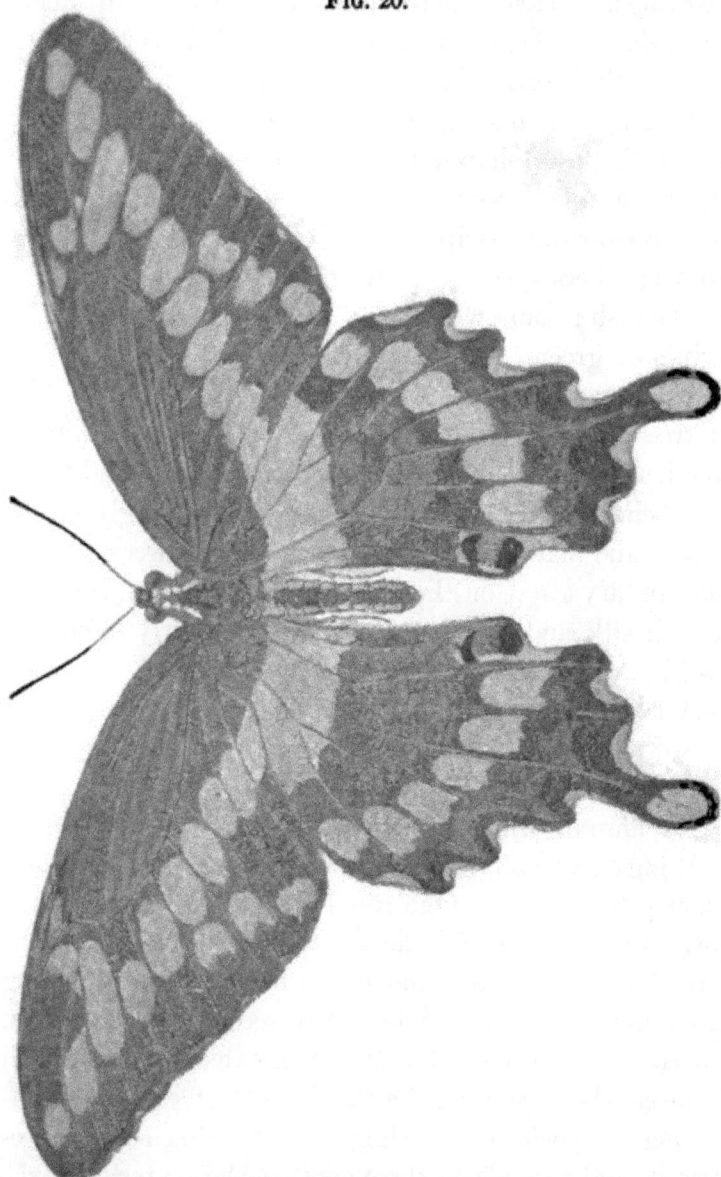

Papilio Cresphontes (natural size).

inner margin just below the ocellus. Ocellus jet-black, with an orange bent bar, and clouded with blue on the basal side. There are some blue clouds inside the yellow row. Tail .4 of an inch long, black, with an ovate yellow spot near the tip.

Most of the ground color of the under side yellow, the blue clouds more distinct, and some orange beyond the discal cell of hind wings and at the anal angle.

Body black above, sides and under parts yellow.

Egg spherical, a little flattened at the base, pale ochre, with sometimes a greenish tinge, at other times inclining to orange.

Fig. 21.

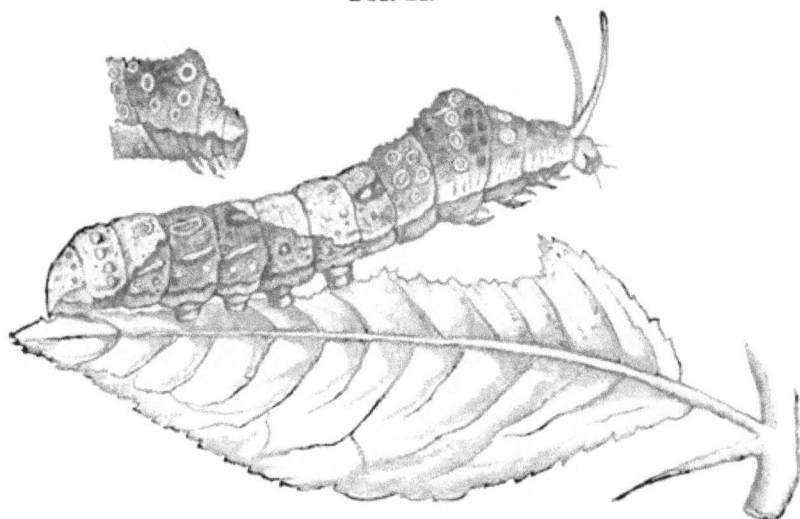

P. Cresphontes, larva.

The young larva is dark brown, beset with tubercles, from which spring short hairs, the sixth and eleventh segments straw color. After the first moult there is but little change, as also after the second. After the third

moult the body becomes shining, the tubercles disappear-
ing, except on joints 2 to 5; thickest through joint 4;
from joint 4 to 5 an abrupt decrease in size, as shown
in the figure.   Head olivaceous, the ridge on joint 2
pale olivaceous, parts of joints 6 and 7 creamy tinged
with olivaceous; the terminal part of the body somewhat enlarged and pearly-whitish on the back, tinged
with olivaceous round the edges; the rest of the body
olivaceous brown.

The mature larva is 1.75 inches long, shaped much
as before the last moult, a prominent ridge extending
across the second segment, along the sides and over the
back of segment 4, this being the highest part.   Inside
this space it is somewhat flattened.   The dark parts are
dark brown; a white band extends from above the head
round to the elevation on joint 4, the lateral portion
being mottled with olive and brown; several white rings
on the elevated ridge, and a few on the dorsum of joint 5. On the dorsum of joints
6 to 8 is a light space extending a little
over on the sides; another similar space
on the posterior part of the body; from
two to four small blue spots on each joint
back of the third.

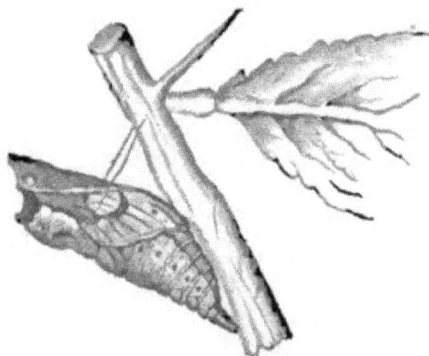

FIG. 22.

P. Cresphontes, pupa.

The chrysalis is 1.5 inches long, some a little shorter;
shaped as in the figure; the abdomen with a subdorsal

row of small tubercles. Color variable. One form gray marked with dark gray and brown, another pale green marked with gray and brown; the latter color mostly on the head and down the ventral part of the thorax.

There are two broods of this insect in a season in this latitude, the larva feeding on prickly-ash, orange, hop-tree (*Ptelea trifoliata*), and *Dictamnus Fraxinella*.

Southern and Western States; Ohio, West Virginia, Michigan, New York, Connecticut.

8. PAPILIO POLYDAMAS, Linn.

Expanse of wings 3.5 inches.

Color of upper surface greenish black; a single row of yellow spots to each wing, nearly parallel with the outer margin. The apex is more produced than in the preceding species, and the row of spots only partly follows the flexures; the row on the hind wings not curved so much as the outer margin, and forming a continuous band but for the black veins. No ocellus or tail.

On the under side the black has a brownish tinge; the yellow spots of the fore wings are repeated except towards the apex, but the yellow spots of the hind wings are absent; but close to the margin are seven red spots, the anal one a bar, the rest more or less figure-3-shaped.

Body black, with a narrow orange stripe on each side, and orange spots on the collar.

Indian River, Florida; Cuba, Mexico.

SUBFAMILY PIERINÆ.

In the United States this subfamily contains all of the Papilionidæ except the genus Papilio. The butterflies do not have the tail to the hind wings, though a few have an angle in the outer margin of these wings; and the inner margin of the hind wings is convex and bent downward so that the two sides form a gutter, in which the abdomen apparently rests. The larvæ are cylindrical, have a few scattered hairs over the body, sometimes a fine short pile also, but lack the scent-organ of the Papilios. Some of the chrysalides, as Pieris and Colias, resemble those of the Papilios except in size, but others are strongly projecting ventrally so as to be nearly triangular.

9. Pieris Ilaire, Godt.

Expanse of wings 2.5 inches.

Wings white; the apex brownish black, the costa and the anterior two-thirds of the outer margin bordered with the same; a very slight border of black in the fringe of the hind wings. This color is not repeated on the under side except along the costa. The basal part of the costa of the hind wings is tinged with dark yellow. Body black, with white hairs.

Indian River, Florida; Texas, Arizona.

10. Pieris Monuste, Linn.

Expanse of wings from 2.5 to 3 inches.

Wings white, costa black, a black border on the outer margin, covering about the outer fourth of the wing at the apex, but narrowing to a point at the posterior angle. This border is serrated on its inner edge, with two or three white rays extending nearly across the border near the apex. Hind wings with a very narrow border composed of triangular spots. Female has a bowed black line on the middle of the fore wings.

The under side has the border less distinct than above, the veins colored, and a shade partly across the middle of the hind wings. In the female the border is more prominent than in the male.

This species, the largest one of the genus with us, is spread over the Gulf portion of the Southern States, where it is known as the Larger Cabbage Butterfly.

According to Professor Riley, the eggs are light yellow, subovoid, with the base applied to the leaf, smooth.

The larva, when full grown, is about 1.6 inches long, lemon-yellow in color, with four longitudinal bands of a purplish shade. Each joint is somewhat spotted with black and covered with sparse delicate bristles.

The chrysalis is pale yellowish marked with blackish, and characterized by two black filamentous spines on the middle of its body.

The food-plants are cabbage, kale, lettuce, turnip; and it has also been found feeding on a species of Cleome and Polanisia.

Southern States, Texas.

11. PIERIS PROTODICE, Bd.—Lec.

Expanse of wings from 1.6 to 1.8 inches.

Summer form, PROTODICE, Bd.—Lec. Male.—Upper surface white, fore wings with a broad black dash or bar across the end of the discal cell (Fig. 23), and a submarginal row of three more or less distinct spots, the last almost or quite touching the hind margin. There are traces of rays running from this row to the outer edge. Hind wings without spots.

FIG. 23.

Pleris Protodice, male (natural size).

On the under side the spots and bars are repeated; the veins of the hind wings are broadly marked with greenish yellow sprinkled with brown scales, and the tips of the fore wings tinged with greenish yellow.

Female (Fig. 24).—The color is the same, and the fore wings have the bar at the end of cell and the subterminal row of spots, but these show a tendency to blend, and the outer margin supports a border of triangles connecting with the subterminal row by rays.

FIG. 24.

P. Protodice, female (natural size).

The hind wings have a zigzag subterminal blackish line, the outer portions sending rays to the margin, where they are somewhat expanded. The base of both wings is more sprinkled with dark scales than in the males. The under side similar to that of the male.

Winter form, VERNALIS, Edw.—This form is smaller than the summer form, and the dark colors are more prominent. The spots of the subterminal row of the fore wings are more inclined to be connected. It expands scarcely 1.6 inches.

Body black, with some white hairs and scales.

The eggs are long, slender, pointed, and deposited singly on the under side of the leaves of its food-plant, often a number on one leaf.

FIG. 25.

P. Protodice, larva and pupa.

The larva when first hatched is of a uniform orange color, with a black head. When full grown it averages 1.15 inches in length and is nearly cylindrical. The most common color is green verging into blue, each joint with six transverse wrinkles. There are four longitudinal yellow lines each equidistant from the other, and each interrupted by a pale blue spot on the first and fourth wrinkles of each joint. There are traces of another substigmatal line. On each wrinkle is a row of various-sized, round, black, piliferous spots, those on wrinkles one and four being largest and most regularly situated; a black hair arising from each spot. Head concolorous with the body, covered with black spots, and usually with a yellow or orange patch each side.

The chrysalis is .65 of an inch long, varying in color, but mostly bluish gray more or less sprinkled with black, with the ridges and prominences edged with buff or flesh color.

This butterfly is usually known as the Southern Cabbage Butterfly, though it is to be found in all parts of the United States, from Canada to the Gulf, and from the Atlantic to the Pacific. Though it has such a wide range, it is to be met with as an injurious insect only in the Southern States and the States bordering on these. In the Northern States *P. Rapæ* is more common, in many places driving out *Protodice.* Where the two occur, the European species is more destructive, as the larvæ of this species not only eat the outer leaves, but may be found boring into the head as well, while the *Protodice* larvæ feed mostly on the outer leaves.

There are several broods during a season, the broods somewhat intermingling, so that larvæ of various stages of growth may be found at any time. It hibernates in the pupa state.

Middle, Southern, and Western States to the Pacific.

12. PIERIS NAPI, Esper.

It has been shown by Mr. Edwards that some one or more forms of this variable species are to be found from Arctic America as far south as California on the west, and Michigan and New England on the east, being mostly represented in the regions farther to the north. As a mere matter of information, the full arrangement of the forms as found in Mr. Edwards's new catalogue is given, with the locality of each, after which those occurring in the Eastern United States will be considered.

PIERIS NAPI, Esper.

>Arctic form, BRYONIÆ, Ochs.—Alaska.
>Var. HULDA, Edw.—Kodiak, Alaska.
>1. Winter form, VENOSA, Scud.—California to British Columbia.
>Aberr. FLAVA, Edw.—California.
>2. Winter form, OLERACEA-HIEMALIS, Harr.
>Var. BOREALIS, Grote.—Labrador, Anticosti.
>Var. FRIGIDA, Scud.—Boreal America.
>Aberr. VIRGINIENSIS, Edw.—New York, Ontario.
>3. 1. Summer form, ACADICA, Edw.—Newfoundland.
>>2. Summer form, *a*. PALLIDA, Scud.—California to British Columbia.
>>>*b*. CASTORIA, Reak.—California to British Columbia.
>>Aberr. FLAVA, Edw.—California.
>>3. Summer form, OLERACEA-ÆSTIVA, Harr.— New England to Michigan; Ontario, Quebec.

Aberr. VIRGINIENSIS, Edw.—Expanse of wings 1.7 inches. Upper side white, less pure than the form *Oleracea*, and much obscured by gray-brown scales, which are scattered over the whole surface, but are dense on apex, costa, and basal half of fore wings, and at base and along the subcostal and median venules of hind wings; a gray patch on costa of hind wings.

Under side white, the venules all bordered with gray-brown, most conspicuously on the median vein of both wings and the branches of this vein on hind wings; shoulder pale orange.

The female expands 1.9 inches; similar to the male, the surface usually still more obscured.

FIG. 26.

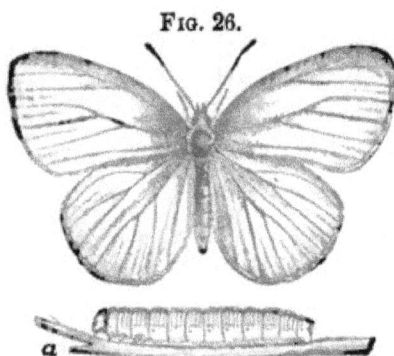

Pieris Napi, winter form, Oleracea-hiemalis:
a, larva.

New York.

Summer form, OLE-RACEA-ÆSTIVA, Harr. This is often of larger size of wings, and the wings are thinner, and purer white on the upper side, than in *hiemalis* (Fig. 26). Also the base is less obscured, and the costa, apex, and outer margin not at all. On the under side it is either white or delicate yellow; the veins of both wings but scantily edged with brown scales, and often not at all over considerable areas.

The females have the basal and apical areas pale gray, and not infrequently there is a trace of the spot of *Napi* on upper median interspace; sometimes also a trace of the second spot, and of the gray bordering to the hind margin of fore wings. The veins beneath are more edged with brown scales than in the male. The shoulders of hind wings are of a *very pale yellow*, and often there is no color at all.

New England to Michigan.

The eggs, represented in Fig. 27, are somewhat pear-shaped, pale greenish white in color, marked with about fifteen sharp longitudinal ridges with cross-lines between. Length about .05 of an inch.

The young larva is of a glassy white, thinly clothed with fine short hairs. As with several other species, the egg-shell from which it emerges forms the first meal of the young larva.

Fig. 27.

The mature larva is about 1.25 inches long, of a pale green color, with a darker dorsal line, the entire surface covered with fine short whitish hairs.

The chrysalis is of a greenish or whitish color finely speckled with black, and shaped much as the other species.

The larva, when ready to pupate, leaves the cabbages and seeks some protected place on the under side of a board or a fence-rail,

P. Napi, form
Oleracea ; egg,
X 27.

where it spins its button and loop of silk and changes to a pupa. This habit is not confined to this species, but is common to the rest of the genus occurring in the eastern United States.

13. PIERIS VIRGINIENSIS, Edw.

This is a form occurring in West Virginia, like *Oleracea*, except that it has no yellow on the under side of the wings. It is single-brooded, producing no summer form, while farther north the aberrant form *Virginiensis* is one of the spring forms of *Oleracea*, and the parent of *Oleracea-æstiva*, a summer form. The preparatory stages are like those of the preceding species, it seeming to be a descendant of one of its forms, probably *Oleracea-æstiva*.

West Virginia.

h   10*

14. Pieris Rapæ, Linn.

Expanse of wings from 1.6 to 1.8 inches.

Upper surface white, the usual form having a brownish- or grayish-black patch across the apex. The male has a submarginal round spot in the first median interspace (see Fig 28), and a somewhat elongated spot on the costa of the hind wings. The females have a second round spot at the same distance

Fig. 28.

Pieris Rapæ, male (natural size).

from the outer margin on the upper side of the submedian vein. The base of the wings is dusted a little with gray scales, more so in the female.

On the under side the fore wings are white, pale yellow towards the apex, and with two black spots in both sexes corresponding to the two on the upper side of

Fig. 29.

P. Rapæ, female (natural size).

the fore wings of the female. Hind wings pale yellow, without marks, but sprinkled with black atoms.

Body black above, white beneath.

Var. Novæ-Angliæ, Scud.—This form occurs, so far as is now known, only in the Eastern States and New York. Ground color of both wings dull sulphur-yellow.

Farther south the winter form, or the one that comes in early spring from chrysalides that have hibernated, tends to pure white on the upper surface. One male in the writer's cabinet has an obscure patch on the apex of the fore wings, and the costal mark of the hind wings about as much obscured, no trace of the spot in the median interspace except what shows through from below. Another male has scarcely a trace of the apical patch, or the costal mark, with perhaps half a dozen scales in the median interspace. On the under side these specimens differ a little from the usual form, both being more suffused with black on the hind wings, the fore wings with scarcely any or no yellow at the apex, and only a few scales in place of the usual dots.

Var. MANNI, Mayer.—This is a pale yellow form, having all the usual markings, but the upper surface pale yellow of a clear type and not ochraceous-tinted. Under side like the others. Found in both sexes.

Georgia; Chicago, Illinois.

The larva of this species feeds on cabbage, turnips, and some other plants. It is not a native of this country, but was introduced from Europe about 1863, since which time it has spread over the most of the United States. It is usually known as the European Cabbage Butterfly.

The eggs are deposited irregularly over the surface of the leaf of the food-plant, mostly on the under side. They are somewhat pear-shaped, flattened at the base, and the apex truncate. In color they are yellowish, marked with twelve longitudinal ribs, crossed by very fine lines between.

The young larva is pale yellow. It first eats the shell

of the egg from which it emerges, then covers a space
with silk, where it rests except when feeding.

When full grown, the larva is about an inch and a
half long, of a pale green color, finely dotted with
black; a pale yellow dorsal stripe,
sometimes indistinct, and a row of
yellow spots along the region of the
stigmata.

Fig. 80.

The chrysalis (Fig. 30, *b*) varies in
color from a dull yellowish green to
an ash-gray, a light gray with numerous black points being the most
common form.

There are probably two broods of
this species in the most northern portions of the United States, in the latitude of Southern Illinois three at
least occur, and it is quite probable
that still farther south there are four or five. Like the
other species, it hibernates in the pupa state.

P. Rapæ: *a*, larva; *b*, pupa.

New England to the Rocky Mountains; south to
Georgia.

15. NATHALIS IOLE, Bd.

Expanse of wings from 1.1 to 1.15 inches.

Upper surface yellow; a large patch of black across
the apex of fore wings, and a stripe of the same along
the hind margin. The fringes and a little along the
outer part of the costa are yellow; and the black along
the hind margin does not quite reach that margin, nor
does it extend to the end of the wing, but bends forward
a little before reaching the posterior angle, where it is

more or less completely separated from the apical patch by yellow.

The hind wings have a black stripe along the basal two-thirds of the costa, the rest of the wing being yellow in the male, except a few black scales on the outer ends of some of the veins; but in the female there is a partial broad, dusky outer border, separated from the black of the costa by a yellow space, the surface having a tinge of yellow.

Fiᴏ. 31.

Nathalis Iole, female.

On the under side the posterior stripe of the fore wings is repeated, in the female somewhat dull, the bent portion being replaced by three dots; in the female these three dots, or spots, form a prominent posterior part of a subterminal row, the posterior stripe wanting or dull. The anterior and outer portion of the fore wings is, in both sexes, washed with orange; the apex and hind wings of the female grayish.

Var. IRENE, Fitch.—This has the under side of the fore wings destitute of a blackish central dot, and of the three black spots near the posterior angle, the posterior one is connected with the posterior stripe; and the base of the wing instead of its outer margin is orange-yellow.

Illinois, Missouri to California, New Mexico, Arizona.

16. ANTHOCHARIS OLYMPIA, Edw.

Expanse of wings 1.25 inches.

Upper surface white, gray at base of wings; a large gray patch at the apex of the fore wings, partially replaced by white. Costal margin slightly specked with black; a black bar at the end of cell. The hind wings have a few black scales at the outer angle and a small wedge-shaped black spot near the base on costa.

Under side white. The fore wings have a small gray subapical patch on costa, nearly covered with green scales, and a faint greenish patch on the outer margin. Discal spot narrow, lunate, enclosing a white streak.

Hind wings crossed by bands of yellow-green on a gray ground. The one near the base is slightly trifid on the costa, the outer one broadly trifid, but running from the outer margin instead of the costa, the middle and outer one joined on the median vein. There is also a spot of the same color between the anterior ends of the second and third.

Body black above, the under side white, the thorax tinged with greenish yellow.

West Virginia, Indiana, Nebraska.

17. ANTHOCHARIS GENUTIA, Fab.

Expanse of wings 1.55 inches.

Upper surface white, with a large orange apical patch, bordered outwardly with black, in which there are seven yellowish-white spots on the edge of the wing. There is a black dot at the end of the discal cell, some specks on the costa, and several somewhat triangular spots on the outer border of both wings.

Under side of hind wings and apex of fore wings pale greenish yellow, the rest of fore wings tinged very slightly with this color; hind wings and apex of fore wings finely netted with black. Discal dot of fore wings reproduced.

Body black above, white below; antennæ annulate

with black and white. Apex of fore wings produced
so that the outer margin is excavated below the apex.

New York to Virginia; Western States, Texas.

18. CALLIDRYAS EUBULE, Linn.

Expanse of wings 2.5 inches.

Upper surface bright lemon-yellow, usually paler on
the internal margin of hind wings; and the male with
paler rays of raised scales extending inward from the
outer margin of the fore wings between the veins, the
anterior five of these rays extending almost to the cell,
the rest triangular. The hind wings have a similar
border, but it is narrower and more continuous. The
male is without spots; the female has a dark brown
spot with a ferruginous centre at the end of the cell
of the fore wings, and the costa and fringe brown, with
brown at the ends of the veins.

The under side of the male is almost a greenish yellow,
with a more or less distinct ferruginous bar at the end
of the cell of the fore wings, and a white or silvery spot
circled with ferruginous on the cell of the hind wings. In
some examples there are no other marks, but in others
there are traces of marks which are more distinct in the
female.

The under surface of the female is greenish yellow,
but little darker than the male, with the costa rosy and
the fringe ferruginous brown. There is a bar at the end
of cell of fore wings composed of five rosy spots circled,
and separated by brown and ferruginous. On the end
of cell of hind wings are two silvery-white spots circled
like the others, and set in a patch of brown and ferruginous scales. On the fore wings beyond the cell are

two rows of elongate, wavy, ferruginous and brown patches, one of these, of three spots, extending from near the apex obliquely inward, and the other, of two spots, submarginal and nearly parallel with the outer margin.

The hind wings are similarly marked, except that the first row has only two spots; there being also six more or less distinct round spots near the base,—two above the cell, one in the cell, two below the cell, and one at the insertion of the wings. There are also three others farther out below the cell. The fore wings are usually sprinkled with fine ferruginous scales.

Thorax black above, the head and prothorax more or less rosy, abdomen yellow; all the under parts yellow. Antennæ rose tipped with ferruginous.

In the larval state this species is said to feed on the species of Cassia.

Southern States to West Virginia and Ohio, Illinois, Iowa, Arizona, Southern California; occasionally in New York and Rhode Island.

19. CALLIDRYAS SENNÆ, Linn.

Expanse of wings from 2 to 2.75 inches.

Male.—Upper surface clear lemon-yellow, the same shade as *C. Eubule.* Like that species, this has an outer border of intervenular spots of the same shade of yellow as the wings, and the scales composing these spots are slightly raised, as though placed over the others, so that in certain lights they seem lighter than the other parts, the remainder of the wing having in the same lights a slight greenish tinge. On the fore wings, beginning at the costa, those in the first and second subcostal interspaces reach almost to the base of these spaces, the same as they do in *Eubule*, the third does not go quite so near, while the fourth does not extend more than a third of the distance from the margin to the cell, in *Eubule* both of these going as near to the cell as do the first and second. The remainder are blunt conical, similar to those of the border of *Eubule*, but a little more blunt. In one small specimen from Indian River, Florida, all of these are more abbreviated than in the above description. On the hind wings the border is a band scarcely sinuous on the inner edge, narrowing towards the anal angle. In the small specimen the border extends along only the anterior half of the margin. Like *Eubule*, there are no colored spots on the upper surface.

Under side a little darker than above, slightly orange-tinted, except along the hind margin of the fore wings. Marked after the pattern of *Eubule*. There are on the fore wings two purplish-brown spots on the end of the cell, the lower twice as large as the upper, elliptical, with an elliptical rosy patch in the centre; the upper with rosy scales on the cross-vein. In addition, there are the usual three series of spots along the outer margin, —the first three parallel with the apical portion of the costa, and situated in the first three subcostal interspaces; the next three in the next three interspaces, extending obliquely inward, the lower spot not quite half-way from the margin to the cell; the third series contains only two spots, lying nearly parallel with the outer margin : these spots vary from a washing to sprinkling of dark brown scales with a few rosy.

Hind wings have one spot at the end of the cell with a central silver spot, and four rows of somewhat scattered spots, all of them a little oblique: the first row consists of

a rosy spot at the base of the wing and a dark one on
each side of the costal vein in line; the second, passing
obliquely through nearly the middle of the cell, contains
three geminate spots; the third, not quite in a straight
line, contains the spot at the end of the cell, which is
composed of several small spots; the fourth, submarginal,
consists of four elongate, irregular patches, the third nearest the margin. Besides these, both wings are sparsely
sprinkled with orange scales. The margin of each wing
is edged with a fine line, with orange-brown points at the
ends of most of the veins.

Female.—About three forms of this sex are to be
met with. One is of the same color above as the male,
with a dark brown round spot at the end of the cell,
nearly divided by a rosy orange line; and a brown
edging along the outer margin of the fore wings, with
small spots at the ends of the veins, these being mere
points on the hind wings, and the edge orange. Another
form is dirty whitish yellow, the discal spot a little
larger; and both wings have a terminal border of quite
prominent, slightly lunate spots, there being four or
five small clusters of scales within the margin in the
subcostal and discal interspaces. Another form is more
like the first, but less clear yellow. A fourth form is
smaller than the others, expanding about two inches.
In this the general color is darker than in the male, with
the hind wings considerably orange-tinted. The spot at
end of cell is fully twice as large as in the first, with the
central spot shorter and broader, the outer border broader
than in the second form, some of the spots approaching
conical, with, on the fore wings, the three series of three
spots each of the under side represented by small patches

of scales.   On the hind wings the two small spots of the under side at the end of the cell show through.

Under side of the same orange-tinted color as in the male, but darker; marks the same, but generally heavier. The first form has the two discal spots at the end of the cell blended, an irregular dark brown outline within in a rosy orange patch, with some silver scales in the centre; the hind wings have a round silver mark at the end of the cell on the cross-vein, and another above and outside, both in a rosy orange patch, with a sprinkling of orange scales, the other spots not heavier than in the male. The second form is dirty whitish yellow as above, as also the third; the discal spot of the fore wings is larger and mostly rosy silver, the terminal border more prominent. The fourth example differs from the others in having the marks much heavier; the spot at the end of the cell of the fore wings is a large silver patch, somewhat divided into four parts; the subterminal spots and the rows on the hind wings are inclined to blend, and the outer margin has a terminal border of rosy scales which shades out into the general color, almost reaching, on the fore wings, the subterminal spots; and the surface is more sprinkled with the rosy scales.

Thorax black above, with whitish hairs; abdomen yellow; head brownish rosy, extending to the tip of palpi, sides rosy; beneath yellow, more or less tinted with orange.

The larva is said to be deep citron-yellow, punctured with black, and a blue transverse line on each segment; abdomen below and feet yellow, with a lateral range of small blue lines above the feet.

Food-plant, Cassia.

Indian River, Florida; Texas, Arizona, Southern Illinois.

20. CALLIDRYAS PHILEA, Linn.

Expanse of wings 3.5 inches.

Female.—Upper surface dark yellow, washed a little with orange along the costa, with a prominent orange border to the hind wings not quite reaching the apex, there being a marginal row of dark brown spots along the outer third of the costa to the apex and round the outer margin of both wings. At the apex these are blended into an apical patch. Discal dot not very prominent. The fore wings have a submarginal row of spots answering to those usually found on the under side.

Under side yellow, heavily sprinkled with ferruginous, more prominent towards the base of the hind wings. Marks similar to those of *C. Sennœ*, but heavy, and the brown has a washing of rose color.

The males are yellow, with a patch of light orange near the anterior margin of the fore wings, nearer the base than the outer margin. Hind wings with the orange border similar to that of the female, except that it lacks the brown spots.

Occasionally in Texas, Illinois, and Wisconsin.

21. CALLIDRYAS AGARITHE, Bd.

Expanse of wings from 2.25 to 3 inches.

Male.—Upper surface clear light orange, a little paler over the inner portion of the hind wings; without spots, but with a terminal border of elevated scales which in certain lights seem to be paler; the border in width nearly one-fourth the length of the fore wing, but about half as wide on the hind wings; crenate on the inner edge on the fore wings.

Under side paler yellow than above, but orange-tinted, scarcely darker than the under side of the male *Sennæ;* nearly without marks. At the end of the cell of the fore wings there is usually a small blackish-brown spot, with or without a few rosy scales, some examples not having either the black or the rosy. There is a more or less distinct oblique stripe of dark scales extending from near the apex to near the hind margin, usually stopping at the lower branch of the median, opposite the lower side of the cell, almost half-way from the outer edge to the cell. The hind wings have at the end of the cell a faint dark brown circle, and one in the interspace above outside the cell; in some examples scarcely a trace of these. Besides these there are traces of a submarginal row of spots, and a row through the end of the cell like *Sennæ,* but they are represented by a few scattered scales or not at all; also some scales in the places along the costa and near the base of the cell, representing an inner row.

Female.—This is more of the color of the dirty yellow form of *Sennæ,* or dirty whitish yellow. At the end of cell of fore wings an elliptical blackish-brown spot; the costa blackish brown, the costal margin sprinkled with this, the apex blackish brown; along the outer margin a series of brown semi-oval spots at the ends of the veins, which are not quite connected, these extending along the hind wings nearly to the anal angle. Extending from the apical patch on the fore wings is an oblique row of six spots separated by the veins, and three smaller ones in the subcostal interspaces. The hind wings have three

11*

submarginal spots, which are not quite so distinct as those on the fore wings.

Under side whitish sprinkled with rosy scales, especially along the outer margin and apex of fore wings and outer margin of hind wings; subterminal row of spots of the upper side repeated, but not the terminal. The spot at the end of the cell is large, long, silvery, surrounded with blackish scales, broken into four or five parts, much as on *Sennæ*.

The hind wings have the silver spots, one on the cross-vein at the end of the cell and the other on the interspace above and outside, these circled with blackish brown and with rosy scales. Besides these there are the same traces of spots that are found on the male, though a little more prominent.

Thorax black, with whitish hairs, abdomen yellow, head and palpi above dark, below concolorous with the wings; antennæ brown, with brownish tip.

Florida, Texas, Kansas, Arizona, occasional in Nebraska.

22. KRICOGONIA LYSIDE, Godt.

Expanse of wings from 1.7 to 1.95 inches.

Upper surface white, in some examples very slightly green-tinted. The fore wings with the basal third bright lemon-yellow, shading a little into the white on the outer edge; the apex yellow-tinted; the base of the wing black-edged. Hind wings uniform white, except a little tinting of yellow at the base.

There are two forms of this, a winter form, LYSIDE, Godt., which has the upper surface as above described; the under side of the fore wings as on the upper surface, except that there is more yellow at the apex, and the costal margin is yellow-tinted. The hind wings uniform pale yellow, heavily sprinkled with whitish or slightly buff-tinted scales, giving the wing a slight grayish cast; a few brown scales on the middle of the cross-vein.

Summer form, TERISSA, Luc., is like the winter form above, except that there is a bronzy black bar about .15 of an inch long on the hind wings from the costa back about two-thirds the distance from the base. The under side of fore wings the same, but the hind wings are more yellow-tinted, lack the whitish scales, and the brown on the cross-vein is more distinct.

Texas; Indian River, Florida.

23. COLIAS CÆSONIA, Stoll.

Expanse of wings from 2.25 to 2.5 inches.

Upper surface yellow, with a broad terminal border, the inner part bent inward along the costa, and making a deep sinus between the second discal and the second median venule. The base of the wing has a heavy shading of black scales, the anterior portion extending half-way across the wing, and from the median vein to the costa. This leaves the yellow portion somewhat resembling a dog's head, the discal dot answering for an eye. Hind wings with a narrow black border, dentate on the inside; an orange discal spot, and a smaller one just outside the cell.

In the female the yellow of the fore wings is more encroached upon by the black basal shading, with a sprinkling of black atoms over the "dog's head," and rays of black between the veins of the hind wings. There is also a slight blue reflection over the "dog's head." Costa, antennæ, and portions of the fringe rosy.

On the under side the discal spots are more prominent than above, and pupilled with silver; a submarginal row

Fig. 82.

Colias Cæsonia, male (natural size).

of dots are black on the fore wings, but red on the hind wings. Ground color of hind wings and apical portion of fore wings dark yellow; a rosy ray from the body outward on the hind wings.

The larva is said to be green, with a lateral white band, punctured with yellow; besides this band, there is on each segment a transverse black band, bordered with yellow. It feeds on the different species of clover.

Southern States, Mississippi Valley, Texas to California; occasional in Minnesota.

24. COLIAS EURYTHEME, Bd.

Expanse of wings from 2 to 2.35 inches.

Varying considerably in color, but the usual form of male orange-yellow, shading to sulphur-yellow on the costa of both wings and on inner margin of hind wings; the base and inner margin sprinkled with black scales. Outer border black, broadest at apex, somewhat irregular on inner edge, extending a little on the costa and hind margin of fore wings; the anterior veins yellow where they cross the black. In width the border is about one-fourth the length of the wing. Discal spot black. On the hind wings the border is narrower, and does not reach the anal angle. Discal spot of hind wings orange, composed of two spots. Both wings have a roseate reflection.

Under side yellow, middle of fore wings tinged with orange. On both wings a subterminal row of dots, the three posterior of the fore wings black, the rest brownish, also two dots on the costa near the apex. Discal spots repeated, the anterior black, with a few light scales; posterior geminate, silvery, annulate with roseate brown or ferruginous; a dash on the costa of the hind wings near the apex, and a rosy spot at the base.

The typical female is of the same general color, a little more yellow along the costa and beyond the discal cell. The border instead of being solid black contains a row of yellow spots, the third from the posterior end on the fore wings subobsolete. The hind wings have the border wider than in the males, and it contains the rudiments of a row of spots. The black scales scattered over the base cover more of the wings than in the male. Under side similar to that of the male.

A white or albino female form is sometimes found, with all the markings as in the yellow form. There are the following seasonal and local variations from the typical form.

i

Winter form, ARIADNE, Edw. This has an expanse of wings in the male of from 1.3 to 1.6 inches; in the female, of from 1.6 to 1.8 inches. The upper surface is of a bright lemon-yellow. On the fore wings an orange patch extends from the hind margin to the median vein or beyond, sometimes very pale, but usually decided and gradually passing into the yellow beyond.

Hind wings sometimes slightly tinted, but more often without orange save the discal spot. Marginal borders narrow, scarcely half as wide as in the form *Keewaydin*.

Under side more greenish yellow than *Keewaydin*, a large double discal spot on the hind wings, silver, annulate with ferruginous, and placed in a patch of pink ferruginous.

In the female the orange on the fore wings is much as in the male, the hind wings greenish yellow much dusted over with black scales. Marginal borders narrow, the border on the fore wings only partly enclosing the submarginal spots, or even without trace of spots, especially on the hind wings.

This form is found only in the Southern States, more distinctly marked in Texas than elsewhere: here the summer form *Eurytheme* flies through the summer, but the forms *Ariadne* and *Keewaydin*, from chrysalides wintered over, take its place in the spring, *Ariadne* being the first one that emerges. In the Northern States *Keewaydin* is the winter form, while in the mountain regions *Keewaydin* and *Eurytheme* are found flying together during the summer.

Winter form, KEEWAYDIN, Edw. This may be known from the typical *Eurytheme* by its smaller size, its duller yellow and less roseate reflection, and both sides more sprinkled with black scales. The costal yellow of the fore wings is broader, encroaching more upon the orange, the latter being deepest near the base. In some specimens there is very little or even no orange on the fore wings, and in the latter case there is no orange on the hind wings except the large discal spot. Those specimens that have considerable orange on the fore wings have the hind wings washed with orange, but not so deep as the fore wings. The orange discal spot is larger than in the form *Eurytheme.*

On the under side the yellow is less of a deep yellow and more of a greenish yellow, resulting from a sprinkling of fine black scales. The discal spot of the fore wings is more or less triangular, white in the centre; the hind wings have one or two discal spots, annulate, with roseate scales, less ferruginous than the typical form.

Summer form, EURYTHEME, Bd.—This is the form first described.

The egg of this species is .06 of an inch long, narrow, fusiform, tapering evenly from the middle to each extremity, the base broad, the summit pointed; ribbed longitudinally, and crossed by numerous striæ. Color buff-white when first deposited, but after one or two days changing to crimson, and near the close of the stage to black.

The young larvæ are cylindrical, of even diameter to the eleventh segment, each segment several times creased, and on the ridges thus formed many black points, from which spring white hairs. Color dark brown or chocolate.

After the first moult the length is .125 of an inch;

shaped as before; body covered with minute black
tubercles, disposed on the ridges so as to form both
longitudinal and transverse rows, each tubercle sending
out a white hair. Color dull green, head ovoid, dark
brown.

After the second moult the length is .28 of an inch.
Color dark green, head as before.

After the third moult the length is .45 of an inch;
cylindrical, long and slender. Color dark green; at
base of body a white stripe, through which runs a crimson line, and under this stripe are black, semicircular or
ovate spots, sometimes seen only on segments 3 to 6, but
usually from 3 to 11, sometimes wanting. Towards the
last of the stage a paler subdorsal line. Tuberculated
and pilose as before.

After the fourth moult the length is .56 of an inch.
Color dark green, but varying, some examples having
the sides only dark, the dorsum yellowish; the subdorsal
stripe sometimes wanting, but usually present. Head
ovoid, yellowish green.

The mature larva is from 1.1 to 1.2 inches long,
cylindrical, each joint as in the early stage several times
creased, and on the ridges thus formed several fine
papillæ, white or black, each supporting a fine short
white hair. Color dark green, at the base of body a
band of pure white, through which runs a bright crimson
line from segments 2 to 11 almost continuously. Beneath this band, from joint 3 to 12, is a large semicircular
or semi-ovate black spot to each joint, the anterior ones
largest. There is a faint white subdorsal line thickened
at the posterior end of each joint so as to present a well-
defined white spot. Above this a line of crimson, broken

on each segment. Beneath these lines, on joints 5 to 10, is a black dot to each joint. Under side, feet and legs pale green ; head ovoid, pale or yellow green.

The chrysalis has the anterior part and the wing-cases dark green, the abdomen yellow-green. There is a light buff stripe on each side of the abdomen from the end of the wing-cases to the extremity, and on the ventral side of this stripe a demi-band of dark brown. Between the stripe and the bend are three black dots, one to each segment, with a submarginal row of black dots on the wing-cases.

Clover forms the food-plant of this species.

Western States to the Pacific; occasionally in the Middle States to Massachusetts.

25. Colias Philodice, Godt.

Expanse of wings from 1.75 to 2.5 inches.

Upper surface of wings sulphur-yellow, with a broad terminal border of black, broader on the fore wings of the female than on those of the male, and containing a submarginal row of yellow spots which are absent in the male. Discal dot of fore wings black, elliptical in the males, oval in the females; on the hind wings orange, usually with a smaller accompanying dot. The antennæ, costa, collar, and fringes are roseate.

Under side about the same color as above, but sprinkled more or less with brown scales, except from the cell to the posterior margin of the fore wing, the winter forms more heavily sprinkled than the summer. Discal spots silvery in the centre, the anterior annulate with black, the posterior brown set in a pinkish-brown patch. There is a submarginal row of dots, the last three on the fore wings black, all the rest brown with pink-brown scales; a roseate spot at the base of hind wings.

A female form occurs with the wings nearly or quite white, and also a black form. The winter forms, or those from hibernating chrysalides, are usually somewhat smaller than the summer forms.

The eggs are pale yellow when first deposited, but change in a few hours to a dark crimson. They are spindle-shaped, attached by one end, ribbed longitudinally, and crossed by numerous striæ. These are deposited on the leaves of clover, Medicago (lucern), buffalo-pea, and some other allied plants. From these a brownish-green larva hatches in six or seven days which is .06 of an inch long, cylindrical, of uniform size from segment 2 to segment 11, then tapering to the last. Color brownish green, each segment creased by four or five transverse creases; each ridge with several black dots on each side, each dot supporting a short whitish clubbed process. Head obovate, dark brown. At first the larva eats little holes in the leaves, but as it grows older it eats the whole leaf from the outside.

After the first moult it is .12 of an inch long; shaped and creased as before; the whole upper surface covered with minute whitish tubercles which are black at their summits, these tubercles forming longitudinal and transverse rows on the ridges. Color dull green; head black.

After the second moult the length is .3 of an inch. Color blue-green, showing a faint whitish lateral stripe; head pale green; tuberculated as before.

After the third moult the length is .7 of an inch; the principal changes are: lateral stripe white and distinct, with usually a red or orange discoloration on the anterior segments, sometimes black lunate spots beneath the stripe.

The mature larva is 1.2 of an inch long; cylindrical, tapering slightly from joint 7 to the anal and from joint 5 to the head; the tubercles and white hairs as before. Color dark green, pale on the under side; in line with the spiracles a white or creamy-white stripe, through the middle of which runs a streak of crimson, broken at the junction of the segments; frequently below this stripe is a series of lunate black spots. Head pale green.

The chrysalis is an inch long, of a yellowish-green color, with a yellow line along each side. From the time that the egg is deposited to the emergence of the butterfly from the chrysalis is about forty days during the warm part of the year, and the number of broods will vary according to the locality. This is not usually considered a very injurious insect, but Professor C. H. Fernald, of Orono, Maine, estimates that these caterpillars often destroy as much as twenty-five per cent. of the entire clover-crop. Their numerous parasites and other enemies serve in a great measure to keep them in check.

Atlantic States to the Mississippi Valley.

26. COLIAS INTERIOR, Scud.

Expanse of wings 2.25 inches.

Professor Fernald says that the males of this species closely resemble those of *C. Philodice*, except that the submarginal row of dots on the under side of the wings is entirely wanting in both sexes, and the terminal black band of the fore wings does not reach the hind margin, and is almost wholly wanting on the hind wings of the females.

Maine.

27. TERIAS NICIPPE, Cram.

Expanse of wings from 1.6 to 1.9 inches.

Upper surface orange; a black terminal band in the males unbroken from the posterior angle of the fore wings to the base of the costa, but broadest at the apex; the basal portion on the costa washed with yellow. On the hind wings the border extends from apex to anal angle, the inner edge irregular. Inner margin of hind wings yellow.

The female has the border of the fore wings broken at the posterior angle, and the anal half of the border to the hind wings is so much suffused with orange as to leave only scattering black scales, most numerous on the veins. There is a narrow black discal spot on the fore wings of both sexes.

This species may vary from the color given above to sulphur-yellow, but orange is the color of most specimens.

Under side of the hind wings canary-yellow, the fore wings yellow along the costa and terminal border, the rest orange. Males with a brown spot on the costa of hind wings two-thirds the length from the body out, and brown scales scattered over the surface; discal spot of fore wings not prominent. Females have a white space towards the outer end of the hind wings enclosed in a subterminal row of brown spots, the first two united and continued obliquely inward nearly across the white space; on the inside three spots in a row, the middle of which is on the end of the cell. Fore wings as in the males.

This species feed on senna, *Cassia Marilandica*, and possibly other species. The eggs are long, narrow, spindle-shaped; the sides marked by about thirty longitudinal ribs without cross-striæ. When first deposited they are greenish yellow, turning red after a few hours.

The young larva is whitish, semi-translucent, a few whitish clubbed appendages to each segment.

The mature larva is about an inch long, cylindrical, thickest through joints 2 and 3. Dorsal surface pale green, the lower part of the sides soft whitish green. Each joint has four or five creases, and on the ridges are small tubercles, which send out short hairs. Along the basal ridge is a whitish stripe, sometimes containing an orange patch to each joint, or there is an orange line the whole length. The chrysalis is long, slen-

der; the ventral side greatly produced, so as to be somewhat triangular. Color of dorsum pale green or whitish green, with a darker line; of wing-cases and ventral side of abdomen, yellow-green, side-ridges cream color, with several brown spots on different parts of the body.

FIG. 33.

Terias Nicippe, pupa.

Pacific to the Gulf of Mexico, Mississippi Valley, Arizona, California; occasional in New England.

28. TERIAS MEXICANA, Bd.

Expanse of wings from 1.6 to 1.9 inches.

Upper surface pale yellow, the anterior half of the hind wings deeper yellow. Fore wings with a broad black ter-

minal border, attenuated on the posterior margin to near
the base, a broad quadrate of yellow in the middle extending two-thirds the distance across the border. Fringe

Fig. 84.

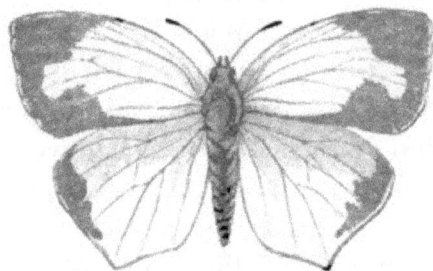

Terias Mexicana, male.

and apical portion of the costa white. The hind wings
have a prominent angle at the end of the first median
venule, a narrow terminal border ending before reaching
this angle, with scarcely a trace of a discal dot. The
female is a little more yellow.

Under side yellow, except the posterior two-thirds of
the fore wings, which are almost white. Discal dots
more prominent than above; a broken brown bar extending nearly across the hind wings from above the
anal angle, and some brown scales scattered over the
surface.

This species seems to be gradually spreading over the
North and East; the habitat as given in Mr. Edwards's
new catalogue being Texas, Arizona, Southern California,
Kansas; occasional in Nebraska, Iowa, Illinois, and Wisconsin, and Ontario, Canada; the last four places having
been added since 1877.

29. TERIAS LISA, Bd.—Lec.

Expanse of wings from 1.15 to 1.45 inches.

Male.—Upper surface yellow, a black terminal border broadest at the apex, narrow at posterior angle, the inner edge of the border somewhat dentate, the costa suffused with black scales. The hind wings have a narrow border also dentate within; a few black scales on the cross-bar of fore wings. Fringe roseate, antennæ and collar black.

Under side uniform yellow, with scattered brown scales, part of those on the hind wings forming a more or less distinct submarginal row of spots; a pinkish or pinkish-brown apical spot to the hind wings.

The female differs from the male in the border of the fore wings not reaching the posterior angle, in that of the hind wings being more or less abbreviated, in the ground color being a duller yellow, and in the base of the fore wings being more densely powdered with blackish.

Specimens occur having the ground color whitish or white.

The larva is green, with four lines along the body, and is said to feed on clover and some other leguminous plants.

Isle of Shoals, Maine; south to the Gulf of Mexico; Western States, Arizona.

30. TERIAS DELIA, Cram.

Expanse of wings from 1.15 to 1.5 inches.

Upper surface citron-yellow, with a broad, black, terminal border, broadest at the apex, and somewhat dentate internally, terminating abruptly before reaching the posterior angle; costa sprinkled with black scales.

A black bar runs along the posterior part of the fore wing, parallel with the hind margin, not reaching the posterior angle, and bordered with darker yellow below.

The hind wings have a somewhat triangular apical patch in line with some indistinct marginal points or rays on the ends of the veins.   Fringes rosy above.

Under side of the fore wings yellow, with the outer margin and apical portion wine-red.   Hind wings tinged with wine-red, and having a transverse undulate, brownish, interrupted band.

On the female the black longitudinal bar is nearly wanting, and the base is sprinkled with blackish.

The larva is green, with a longitudinal white line above the feet, and is said to feed on clover, Cassia, and perhaps other allied plants.

Gulf States.

31. TERIAS JUCUNDA, Bd.—Lec.

Expanse of wings from 1.2 to 1.4 inches.

Closely related to the preceding, but is white on the under side of the hind wings, and without pink or wine color on the under side of the fore wings.   The white is sprinkled over with gray scales.

The female is paler, marked like the female of *T. Delia*, the fore wings powdered with blackish.   Under side like the male.

Gulf States.

FAMILY NYMPHALIDÆ.

These may be known by their ample wings, slender antennæ, the knob slender or not, and by having in both sexes, with the exception of the genus Libythea, only four feet adapted for walking. The front pair of legs are present, but have no developed tarsi, being mere lappets placed against the prothorax; the second pair are directed forward, and the third or hind pair backward. The larvæ are more or less hairy, or covered with more or less branching spines; the head is more or less bilobed, the apex of these lobes often supporting branching spines. The chrysalides are naked, often very irregular in shape, and attached to a button of silk by the hooks of the cremaster alone. The family is represented in the United States by five subfamilies,—Heliconinæ, Danainæ, Nymphalinæ, Satyrinæ, and Libytheinæ

SUBFAMILY HELICONINÆ.

In this the wings are long, rather narrow, with a slender body and antennæ. It is represented by but one species, *Heliconia Charitonia*, where the characters of imago, larva, and chrysalis may be found.

32. HELICONIA CHARITONIA, Linn.

Expanse of wings from 2.5 to 4 inches.

Wings long and narrow. Upper surface black, banded with lemon-yellow, as in Fig. 35; three of these on the fore wings and two on the hind wings. The outer one on the fore wings is obliquely transverse before the apex, the second nearly parallel just outside the cell, the third extends from the base on both sides of the median vein to the third venule, from which it bends obliquely outward, reaching the margin in a dot.

The two basal bands of the hind wings form a straight line when the wings are spread; below this is a line of

Fig. 85.

Heliconia Charitonia (natural size).

dots, the outer end bending round so as to form a submarginal row from about the middle of the outer margin to the apex. There are a few marginal dots at the anal angle, and usually two or three red dots at the base.

Under side dull black, with the yellow lines and dots repeated, though paler. The costa of the fore wings with red at the base, three red dots on the base of the hind wings, and two below the first band.

The egg is described by Mr. Edwards as cylindrical, one-half higher than broad, flat at base, tapering very slightly from base to about three-fourths the length, then conoidal, the top flattened and a little depressed. Marked by fourteen longitudinal ridges crossed by low horizontal ridges. Color yellow.

The young larvæ are cylindrical, tapering slightly from about the seventh segment; marked by four principal rows of flattened tubercles and two rows of smaller ones. Color pale reddish brown.

After the first moult the color is light brown, changing as the stage proceeds to greenish white mottled with brown; armed with six rows of spines, which are short, slender, tapering, and black, with a few short black bristles on the sides. The truncated head is a little depressed in the middle, and each vertex armed with a short tapering black process thinly beset with bristles.

There are but few changes during the next intervals, except in size, and in the color gradually becoming whiter.

The mature larva is from 1.25 to 1.5 inches long, cylindrical, armed as after the first moult. Color dead-white, with no gloss, smooth, no hairs, and spotted with black or brown.

The chrysalis is very irregular in shape, two leaf-like appendages extending from the head. Color brown, marked with varying shades of the same, and some gray or whitish.

This insect feeds on the passion-flower; and there are many interesting things connected with its life and habits.

Florida to South Carolina.

SUBFAMILY DANAINÆ.

In this group the head is broad, the palpi far apart. The wings are ample, the discal cell of the fore wings open, but that of the hind wings closed, or with a vein across the outer end of it. The larvæ are cylindrical, banded transversely, two fleshy appendages from the top of the joints near the end. The chrysalides are well represented in Fig. 39.

33. Danais Archippus, Fab.

Expanse of wings from 3.75 to 4.5 inches.

Upper surface tawny red or fulvous, with the veins heavily marked with black, a black terminal border containing two rows of white spots, and a complete and a partial row of white or lighter fulvous spots in a black space beyond the cell of the fore wings. The males have

Fig. 36.

Danais Archippus, male (natural size).

a black spot beside the second median venule, near the middle of the hind wings.

The under side is paler than above, especially the hind wings, and the white spots are more prominent.

Body black, with white spots.

Usually the larva of this species is to be found on the different species of milk-weed (Asclepias), but it feeds on other plants of the order as well.

When first deposited, the eggs are white, but in two or

three days they turn yellow, and just before hatching they change to dull gray. They are somewhat conical in form, and marked by about twenty-five longitudinal

Fig. 37.

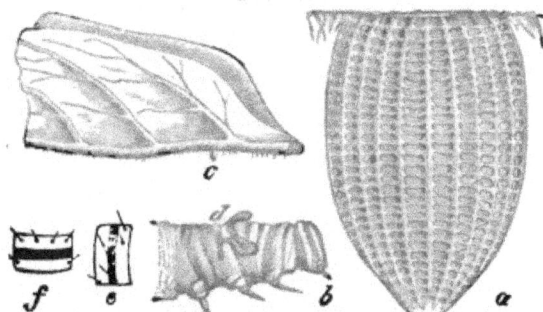

Danais Archippus: *a*, egg, ✕ 30; *c*, natural size.

ribs, with about the same number of transverse ridges, as shown in Fig. 37.

The young larva, which hatches from this in about a week, is yellowish white, with a large black head. It first eats the egg-shell, after which it eats the leaves.

Fig. 38.

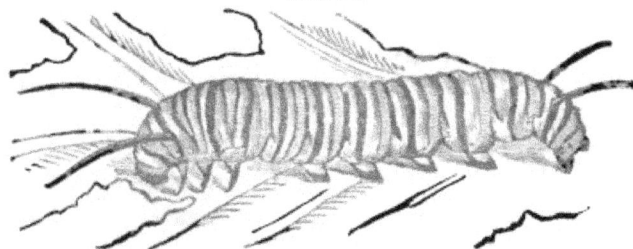

Danais Archippus, mature larva (natural size).

The mature larva is about 1.75 inches long; the head yellowish, marked by two triangular black stripes. The body above is marked with transverse stripes of black, yellow, and white, as shown in Fig. 38. Joint 3 supports

two long, black, fleshy horns which point forward ; joint
11 has a similar pair which point backward, but they are
shorter. Under side black, with green-
ish between the joints.

FIG. 39.

Danais Archippus,
chrysalis.

The chrysalis (Fig. 39) is about an
inch long, color bright green dotted
with gold, and with a band of golden
dots extending more than half-way
round the body above the middle.
The band is shaded with black, and
the cremaster is black. There are
two or more broods in a season, and
it hibernates in the butterfly state.

United States generally.

34. DANAIS BERENICE, Cram.

Expanse of wings from 2.75 to 3.5 inches.

Upper surface reddish chocolate-brown, with a black
terminal border containing two partial rows of white
dots on the fore wings, but the dots are obliterate on the
hind wings. The fore wings have two oblique rows of
white spots beyond the cell, the inner one crossing the
end of the cell, and a few dots forming a submarginal
row. The males have a black spot beside the second
median venule of the hind wings.

The under side is similar to the upper, except that the
terminal border contains two full rows of white spots,
and the veins of the hind wings are heavily marked
with black edged with gray.

The larva is "whitish violet, with transverse stripes
of a deeper color ; a transverse band of reddish brown
on each ring, divided in its length by a narrow yellow

band. Along the feet a longitudinal band of yellow-citron. Long, fleshy processes of brown-purple, disposed in pairs on the second, fifth, and eleventh rings."

The chrysalis is similar to that of *D. Archippus;* green, with golden points on the anterior side, and a semicircle of the same color on the dorsal side, a little beyond the middle, separated from a blue band by a row of small black dots. The larva feeds on Nerium and Asclepias.

Southern States, Colorado, New Mexico, Arizona.

SUBFAMILY NYMPHALINÆ.

The palpi are approximate, more or less porrect; the discal cells generally open, and the veins of the fore wings not dilated at the base. The wings are various, but none of ours have as narrow wings as the Heliconinæ. The larvæ are cylindrical, and furnished with several (usually seven) rows of more or less branching spines or tubercles. The chrysalides vary from nearly cylindrical to considerably depressed on the dorsal side just back of the thorax, as in Figs. 41, 47, 51, etc.

35. COLÆNIS JULIA, Fab.

Expanse of wings 3.1 inches.

Upper surface clear reddish fulvous, the prominent veins of the fore wings narrowly black; the fore wings with a narrow terminal black border without spots, but the border on the hind wings twice as broad as the one on the fore wings, and containing two more or less complete rows of narrow fulvous spots. Costa black, with a narrow fulvous line between the black costal and subcostal veins. Outer third of costal region and round the apex has the black border a little widened. Above the outer end of the cell begins a curved black stripe which runs from the subcostal vein across the upper part of the end of the cell, and along the first median venule to the outer border; a spur from the border above this stripe extends inward one-third the distance to the cell. The subcostal vein beyond the cell is fulvous to the black at the apex.

Under side pale fulvous brown, paler on the outer third, and somewhat clouded. The hind wings have at the base two small white spots annulate with black, and a fulvous spot. At the posterior angle of the fore wings are two geminate whitish spots in black, at the anal angle two pairs of these spots, and one pair at the apex: these spots form part of two indistinct pale fulvous terminal lines. Fore wings long and narrow, the hind margin not more than half the length of the costa; hind wings triangular.

Southern Florida, Texas.

36. AGRAULIS VANILLÆ, Linn.

Expanse of wings from 2.25 to 3 inches.

Upper surface rich reddish or yellowish fulvous, the veins of the fore wings black on the outer two-thirds of the wing, the black enlarged at the ends of the median venules and submedian vein. There are three white spots in the cell of the fore wings, each set in a black patch, one at the end and two in the middle; and three black spots between the submedian vein and the median venules. Hind wings with an outer border of black containing circular fulvous spots between the veins, and three black spots, one in the cell and two submarginal.

Under side of hind wings and apical portion of fore wings yellowish brown streaked a little with yellow, the rest of fore wings fulvous. There are twenty-three or twenty-four silver spots edged with black on each of the hind wings, and about ten on the apex and outer margin of the fore wings; those on the hind wings and apex of fore wings mostly large.

The eggs are conoidal, truncated, the top a little arched; the sides more or less convex, marked by fourteen ribs from base to top, and crossed by eleven tiers of striæ; the spaces between the ribs are quadrangular, the spaces at the summit hexagonal.

The young larva is cylindrical, thickest at joint 4, tapering slightly to the anal extremity. Color brownish orange, glossy; on each side of the dorsal line, on each joint after the second, is a row of conical, pale black tubercles, and two similar rows on each side forming transverse rows of six tubercles, from the top of each of which springs a short black hair. On joint 2 is a black dorsal collar with fine tubercles. Head brown.

After the first moult the color is about the same, but after moulting again it is more of a dark or red brown, with a subdorsal greenish-brown band, and head black. After the third moult the color changes to dark glossy orange, with the dorsal stripe olive-brown and a subdorsal of the same, and the lower part of the body olive-brown.

The mature larva is 1.5 inches long, of a red-orange color, with a broad dorsal line of greenish black, and a broad slate-black band outside this reaching to the first lateral, except a narrow stripe of the ground color. Base slate-black, orange through the region of the spiracles.

Each segment is furnished with six long, tapering black spines, blunt at the top, from each of which springs a bristle. Feet and legs black ; head ovoid, deeply cleft, with high conical vertices, on each of which stands a stout, spinous, recurved process.

The chrysalis is a little more than an inch long, slender, the thorax much compressed, the wing-cases very prominent, forming a narrow carinated hunch, which rounds abruptly on posterior end. Colors variable, some specimens buff with greenish markings, or on the abdomen greenish brown ; some black, the wing-cases and anterior parts mottled in light and dark black ; some with the anterior parts pink-tinted mottled with greenish black.

The larva of this beautiful insect feeds on the passion-flower. It is found in the Southern States ; Arizona, California, and occasionally as far north as Coalburgh, West Virginia ; Cape May, New Jersey ; Philadelphia, Pennsylvania.

37. ARGYNNIS IDALIA, Drury.

Expanse of wings from 2.75 to 3.6 inches.

Male.—Upper surface of fore wings fulvous, black along the costa, with a black outer border which is a little wider than the costal border ; base and hind margin brown. In the cell are three black bars, at the end another bar with an open § united to it enclosing a fulvous spot. Beyond the cell runs a transverse zigzag line, a submarginal row of black dots, and next the border a row of black crescents. On the costa, instead of a subterminal spot there is a black patch, with another between this and the zigzag line.

Hind wings black, with violet reflections; the base of the wing washed with fulvous. There is a black spot in the cell, an irregular row of yellowish spots beyond the cell, and a marginal row of fulvous spots. Fringes alternate spots of black and white. Under side of fore wings fulvous, white along the costa, a marginal row of silver spots enclosed in black crescents, and some silver on the costa and near the apex. The black of the upper side repeated. Under side of hind wings yellowish brown, with twenty-nine silver spots and patches, besides some silver shading.

The female differs from the male in being larger, in the terminal band of the fore wings being broader and containing a row of white spots, with six more white spots near the apex, and in the outer row of spots on the hind wings being of the same color as the inner.

An aberrant form, ASHTAROTH, more suffused than the typical form, is sometimes found.

Mr. Edwards describes the egg as conoidal, truncated, rounded at the base, the sides well rounded, depressed at the summit, marked vertically by about eighteen ribs, half of which extend to the summit, and between these equidistant transverse slightly-raised striæ.

In about twenty-five days the larva hatches from this. It is cylindrical, somewhat thickest in the middle. Color pale yellow-brown, translucent; each segment from 3 to 12 marked by a transverse row of eight elongate tubercular dark spots, the whole forming eight longitudinal rows; one or two long, black, curved hairs arising from each tubercle. Head bilobed, the vertices rounded.

After the first moult the color becomes cinereous, mottled and striped with brown; a macular stripe along

the dorsal rows of spines, and another just outside the
first laterals. The spines from six rows are long, fleshy,
black, each beset with short black hairs. Head black.

After the second moult the larva is mottled and striped
with light and dark cinereous, the spines longer, each dull
yellow at base.

After the third moult the dark portions become black,
and the light a dirty white, and the dorsum has a white
stripe with a central black line; at the juncture of several
segments a transverse white stripe, on which are short
black lines. Each segment is crossed longitudinally by
black stripes, interrupted by the spines, with a wedge-
shaped mark between the spines. Head light brown.

The color after the fourth moult is buff, with the markings much as before. The larva moults five times before
reaching maturity, when it is 1.75 inches long, velvety
black, banded and striped with ochrey yellow changing
to dull orange or red, and furnished with six rows of
tapering, fleshy spines, each of which has several small
black bristles. Two of the rows along the back are
silvery white, with black tips, those at the end of the
rows somewhat smaller. The spines of the rows along
the sides are smaller, and yellowish or orange at the base.
The head is reddish above and black beneath.

The chrysalis, to which the larva changes in some sheltered place, is 1.1 inches long, and shaped as in allied
species. The color is brown and yellow over the abdomen, the mesonotum pinkish brown, the wing-cases
brown, pink-tinted, with dark brown and black patches
over the body.

This species, like others of the genus, feeds on violets
in the larval state. In the North it is single-brooded,

but in the southern part of its range there are two broods in a season. It occurs from Maine to Nebraska, New Jersey, Arkansas.

38. Argynnis Diana, Cram.

Expanse of wings from 3.25 to 4 inches.

Male.—Upper surface from the base to beyond the middle of the wings dark velvet-brown, the rest of the wings deep orange, forming a wide band, crenate next to the brown, and with a brown shading along the veins almost to the edge, and a brown line near the margin. Inside this line are two rows of brown dots more or less distinct, one submarginal, the other next to the brown space.

Under side of fore wings black at base, beyond which are the zigzag and other markings found on the under side of *A. Cybele* and other related species, the color between these markings that of the outer part of the wing above, but somewhat suffused with black. Outer part similar to that above, but paler.

Hind wings with the basal two thirds of a leaf-brown color, the outer part same as above, without spots. Between these parts is an edging of black, more or less covered with silver scales, terminating at each margin in a triangular silver spot. Between the costal and sub-costal veins is a silver crescent edged internally with black; some silver scales at the juncture of these veins, and in the cell. Just within the margin rests a continuous band of silver crescents.

The female has the upper part blue- or green-black, the outer third of fore wing a little paler, with three rows of blue or whitish spots; the inner row reinforced by three more at the end of the cell.   The hind wings have the style of the marks more like that of the males.

On the under side the basal two-thirds of the fore wings are marked like the males, but the colors are black and pale blue ; the apical portion dark brown, an apical whitish spot edged within with silver, a row of whitish spots near the margin, another answering to the inner one above, and between these several light bars tinged a little with pale buff.   Hind wings with the basal two-thirds dark brown, the outer portion with the veins brown, and the part between the veins black washed with brown.   The silver marking the same as in the males.

The egg is conoidal, truncated, depressed at the summit, marked vertically by about eighteen prominent, slightly wavy ribs, eight of which extend from base to summit and form there a serrated vein or crown, the ribs crossed by about twelve transverse striæ.

The young larva is about .05 of an inch long, cylindrical, greenish brown, with rows of tuberculated darker spots, from each of which grows a black hair ; head brown.

The mature larva is velvety black, the body armed with six rows of long fleshy spines which radiate from it like spokes, and from each of which proceed several short black bristles.   The length of most of these spines is .2 of an inch, but the two on the top of the second segment are .3 of an inch long and bend forward over the head. The base of the spines is deep orange or fulvous.   Between each pair of dorsals are two whitish dots placed transversely.   The head is brown in front and fulvous behind.   When full grown it is 2.5 inches long.

The chrysalis is cylindrical, with a depression on the dorsal portion near the anterior part, and several slight elevations on the anterior part. Color brown, marked with different shades of the same.

Food-plant, violets.

West Virginia to Georgia, Southern Ohio to Illinois, Kentucky, Tennessee, Arkansas.

39. ARGYNNIS CYBELE, Fab.

Expanse of wings from 3 to 3.5 inches.

Upper surface fulvous or yellowish fulvous, the base of the wing to the end of the cell, and below this to the zigzag line, yellowish brown, there being more of the yellow in the male. In the cell of the fore wings are the usual five black bars, all but the fourth bent outward in the lower half, the two outer united above. Beyond the cell are the usual zigzag black line and the subterminal row of dots, the middle ones of the fore wings the largest. Just within the outer margin is a terminal narrow line, and within this, and on the fore wings, with their points resting on this line, is a series of crescents; the fore wings edged with the same color. The cell of the hind wings with three more or less distinct bars.

Under side of fore wings pale yellowish brown, the apical space yellowish and enclosing a bright brown costal patch; the lines and dots the same as above, but near the apex more brown. The apical five or six of the spaces enclosed within the submarginal crescents are wholly or partly silver, with three silver patches within this line.

The hind wings have the basal two-thirds reddish brown more or less mottled with yellow, the outer boundary of this color a row of seven silver spots. Outer margin brown, fading into yellow at the anal angle, and within this another row of seven large silver spots rounded within and edged with brown. Between these two rows is a bright yellow band without spots. In all there are twenty-four or twenty-five silver spots to each wing.

Like both the preceding species, this feeds on violets. The egg is conoidal, truncated, broad at base, the sides moderately rounded, depressed at the summit; marked by eighteen longitudinal ridges, half of which reach the summit, with transverse striæ between the ribs.

The young larva is like that of *A. Diana*. The mature larva is from 1.8 to 2 inches long. Color velvety black, the under side chocolate-brown. As in *A. Diana*, there are six rows of slender black spines which are reddish yellow at the base, and beset with many short black bristles. Between each dorsal pair of spines on the joints from 3 to 11 are two gray transverse dots. The spines of the second joint are wholly black, and directed forward, but they are not longer than the others. Head small, subcordate, the front flattened and finely tuberculated, the back rounded, the vertices having on the anterior side of each a small black process. Color of front dull dark brown, of back reddish yellow.

The shape of the chrysalis is similar to that of *A. Diana*. The color is variable, sometimes glossy dark brown, with fine mottlings of reddish orange not distinct, or dark brown mottled with drab, or dark brown mottled with light brown.

Atlantic and Western States to Nebraska.

40. ARGYNNIS APHRODITE, Fab.

Expanse of wings from 2.5 to 3 inches.

Upper surface of wings bright reddish fulvous; the basal third of both wings washed with cinnamon-brown. The black markings, similar to those of *A. Cybele*, but not quite so heavy, are shown in **Fig. 40**.

The black bars forming the median zigzag line are often not connected by black on the veins, so that they

FIG. 40.

Argynnis Aphrodite (natural size).

form a broken line. The two lines at the outer margin of the female are more or less blended, and the two are present on the hind wings of both sexes. The under side of the fore wings is pale reddish fulvous, the apical portion and along the costa buff, with pale brown markings; six marginal and three submarginal silver spots. The hind wings are cinnamon-brown, marked as in *A. Cybele*, but. the submarginal yellow band is narrower, spotted with brown, almost or quite obliterate on its extremities. The silver spots are smaller than in *A. Cybele*, and are more or less edged with black.

14

This may be known from *Cybele* by its smaller size, by its being less brown on the base of the wings on the upper side, and by the submarginal band on under side of hind wings being narrower and spotted with brown.

The preparatory stages are almost identical with those of *A. Cybele,* though the larvæ are a little smaller; and this also feeds on violets.

It is found in the Northern, Middle, and Western States to Tennessee; also in Nebraska, Montana, and Kansas.

41. ARGYNNIS ALCESTIS, Edw.

Expanse of wings from 2.5 to 3 inches.

This closely resembles the preceding, the upper surface being fulvous, in the female a little more yellowish than in *A. Aphrodite,* and the brown at the base a little wider, but narrower than in *A. Cybele,* extending on the fore wings from the inner bar of the cell outward to the lower end of the zigzag line; this line being continuous instead of broken on the fore wings of the female. On the hind wings of the female there is a round black spot in the cell nearer the base than the other usual marks.

Under side of fore wings of the male reddish fulvous, the apex cinnamon-brown, with the usual black and silver spots. Under side of hind wings uniform brown, without the submarginal yellow band, the silver spots the same as in the preceding species. The under side of the female is the same, except that the color on the apex of the fore wings and the whole of the hind wings is rather dark reddish brown, with sometimes a few yellow scales near the central silver spots. The female has nine silver spots on each fore wing and twenty-five on each hind wing, and the black marks of both sexes are heavier than in *A. Aphrodite*. This species may readily be known from the preceding by the absence of the yellow submarginal band on the under side of the hind wings.

The eggs are described as conoidal, truncated, not so broad at the base as *Idalia*, the sides less rounded; depressed at the summit; marked vertically by about eighteen prominent, slightly wavy ribs, half of which reach the summit; and crossed by transverse striæ.

The young larva is translucent greenish brown, each joint from 3 to 12 marked by eight rows of tubercular dark spots, from each of which arises a long, black, clubbed hair, which is curved forward. On the second segment is a blackish dorsal patch, with two small spots on each side, all furnished with hairs.

The color after the first moult is yellow-green mottled with brown on the dorsum; as in the other species, six rows of spines; the dorsals begin at joint 2 and run to 13, the laterals begin at 5 and stop at 12 and 13. The spines are long, tapering, black, and beset with many short and fine bristles. Head subcordate, black.

After the second moult the color is black-brown, the sides less dark than the dorsum. The tubercles of the dorsal spines are buff on the outside; the first laterals have black tubercles, the second buff; the intermediate tubercles on anterior segments are yellow. After the next moult the color is velvety black with a brown tint, with the buff changed to dull yellow and a little of it on the first laterals. The head has the front shining black; the back is yellow. After the fourth moult the

yellow is orange or reddish yellow, and the lower lateral
spines are of this color half-way to the tip.

The mature larva is about 1.5 inches long, cylindrical,
velvety black. The six rows of spines are of about
equal length, the dorsals about .15 of an inch, those on
the top of the second segment directed forward, and all
are beset with short black bristles. Those on the dorsal
rows are translucent brown at base, except on joints 3
and 4, where they are dull yellow; all of the two lateral
and the intermediate rows are dull yellow from the base
half-way to the top; tops of all the spines black. Head
subcordate, deeply cleft, flattened in front, on each vertex
a small conical process. It moults five times in coming
to maturity.

The chrysalis is of the same shape as that of *Diana*,
the color varying. Some are red-brown irregularly
mottled with black, others are drab and black.

Like the others, the food-plant is violets.

Michigan, Illinois, Iowa, Montana, Colorado.

42. Argynnis Atlantis, Edw.

Expanse of wings about 2.5 inches.

Upper surface fulvous, obscured by brown scales on
the fore wings from the second bar in the cell obliquely
to below the median zigzag line on the hind margin;
the two marginal lines of both wings so blended that
not much of the ground color is left. Marks as in *A.
Aphrodite*.

The under side of fore wings is reddish fulvous, the
costa and apex light buff, the apical patch and outer
margin brown, with the usual apical silver spots. The
hind wings are dark red-brown, much mottled with

greenish gray or drab, the submarginal band pale yellow, usually pure from margin to margin. Silver spots the same as in *Aphrodite.*

This species resembles *A. Aphrodite,* but may be known by its smaller size, and by its being more brown at the base of the wings above, and having a darker color on the under side of the hind wings.

FIG. 41.

The early stages are almost the same as those of *A. Cybele* and *Aphrodite,* and the food-plants are violets. The pupa, or chrysalis, is represented in outline in Fig. 41. It has been found in New England, New York, and Iowa.

Argynnis
Atlantis,
pupa.

43. ARGYNNIS MYRINA, Cram.

Expanse of wings from 1.7 to 1.85 inches.

Upper surface yellowish fulvous, less than the basal fourth of the wing dusky brown. In the cell of the fore wings are the usual five bars, the second and third united, but not the fourth and fifth, the fourth an open 3. Beyond the cell the usual black zigzag line; and below the cell under the double bar a longitudinal dash, with projections towards the cell, the inner running to the base of the wing. Outer margin black, inside this a line composed of crescents, with the usual submarginal row of black dots, the whole more or less blended at the apex, so that the marginal line and the row of crescents form a band containing a row of fulvous spots.

Hind wings with the margin and row of black spots as in the fore wings, except that they are less prominent anteriorly. Within the median zigzag line is another crossing the end of the cell, where it sends out a short

l   14*

spur, the cell containing two more or less distinct round spots.

The under side of the fore wings is fulvous, the apical portion yellow, the markings on that and the outer margin rusty brown; the black marks much as they are above; a marginal row of silver crescents, and three subapical.

The hind wings rusty brown mottled with patches of yellow, mostly through the middle and outer portions; a marginal row of seven and a submarginal row of eight silver spots; between the two rows a row of black-brown dots. Inside the second row about eight more silver spots, one in the cell pupilled with black.

Fig. 42.

Egg of Argynnis Myrina,
× 24.

The eggs are pale green, shaped somewhat like the frustum of a cone, and marked with about fourteen longitudinal ribs and fine transverse striæ (Fig. 42).

The young larvæ are pale green, with a brownish-black head. Brown patches nearly cover joints 5, 7, 9, and 11. Black hairs arise from tubercles on all the joints and curve forward. In passing from the young to the mature larvæ they moult four times. The mature larvæ are an inch long, ashy brown mottled with velvety black, with six rows of fleshy spines beset with black bristles, those on the second segment three and a half times as long as the others and pointing forward. Head bronze-colored.

The chrysalis is .6 of an inch long; light brown streaked with darker, and armed with two rows of sharp conical tubercles on the back. The perfect insect flies

from June to July, and is found from New England to
Montana, and in Colorado.

It feeds on violets.

44. Argynnis Montinus, Scud.

Expanse of wings 1.75 inches.

Upper surface rich reddish fulvous, much the color of
A. Bellona, marked with the usual zigzag line beyond
the middle and the row of round black spots; the submarginal row of black lunules and the black terminal
edge somewhat suffused on the fore wings, so that there
is but little clear fulvous between the edge and the
lunules.   The cell of the fore wings is marked with four
marks,—three black bars almost straight, and an elongate O : the latter is the second from the base, and may
be considered as formed of two bars, making the number
five, as in other species.   Below the cell there is an open
V, the point turned outward.   The basal portion of both
wings is suffused with black, extending out somewhat
along the posterior and internal margins.

The under side of the hind wings and the apical portion of the fore wings are deep cinnamon-red, the rest is
ochraceous fulvous, the markings of the fore wings faintly
repeated.   Hind wings with the median broken line repeated less distinctly than above, obsolete opposite the
cell, and partially so between this and the costa.   Costa
black, more or less bordered within by ochre scales.   In
place of the round spots of the upper side there is a series
of ferruginous spots, some indistinct and others with a
few black scales.   Between this and the median line is a
broken line or shade of salmon scales, not very clear,
and outside this the round spots are patches of ochre

scales. Just within the outer border is a series of spots which are white rather than silver, the anal and the two next the costa rather distinct, the rest not very prominent. At the end of the cell is a curved black line, and below this are two more, each bordered on the outside with white or whitish. Above the cell is a straight black line, outside of which is a white patch. Near the base are three white spots with some black scales, and a black spot with a few white scales near the end of the cell. The males and females do not differ.

Found on the lower half of the barren summits of the White Mountains, New Hampshire, during July and August.

45. Argynnis Bellona, Fab.

Expanse of wings about 1.6 inches.

Upper side fulvous, the marks on the wing inside the terminal border similar to those of *A. Myrina*, but the dusky brown covers fully the basal fourth of the wings. In the male the edges of the wings are scarcely black, but in the female they are distinctly so. Inside the terminal edging is a row of oval spots instead of crescents, with some suffusion between this and the edge.

The under side of the fore wings is fulvous, with the apical portion rusty brown, the apex yellow, the brown tinged with purple. Hind wings rusty brown tinged with yellow in the middle and outer two fifths, and a costal patch washed with whitish purple, this portion containing two rows of dark spots. There are no silver spots. The yellow on the middle of the under side of the hind wings is in scattered scales, not gathered together enough to form a band or spots.

According to Professor Fernald, the eggs of this species are similar to those of *A. Myrina* in form, size, color, and markings, and it closely resembles that species in all the early stages. The mature larva, however, does not have the spines on the second segment lengthened.

Food-plant, violets.

Northern United States, Colorado.

46. EUPTOIETA CLAUDIA, Cram.

Expanse of wings from 1.75 to 2.75 inches.

Upper surface fulvous, a paler band crossing both wings near the middle, bordered on the inside by a zig-

FIG. 43.

Euptoieta Claudia (natural size).

zag black line similar to that in the species of Argynnis. From this pale space to the base the wings are somewhat duller fulvous and a little powdered with black scales. Beyond the central pale band are two transverse lines, with a row of round spots between them, the edge of the wing black; all these lines are connected by black along the veins. The cell of the fore wings contains three black bars, the two outer united at the ends and enclosing a pale space. Below the cell is a bar bent outward in the middle.

Under side of fore wings fulvous to the zigzag line, with discal pale spot. The outer half of the wing is pale, with a little submarginal reddish wash below the apex, and a large gray triangle on the costa. A brown spot near the posterior angle sends a marginal streak towards the apex.

The hind wings are pale brown in the basal half, streaked with white along the veins, and with transverse striæ of darker brown. Beyond the middle they are whitish, shading off into the same brown as the base, with more or less whitish along the margin, the anal portion of the outer half being nearly as dark as the base, while the costal portion is almost white. There are about three indistinct submarginal ocelli.

The egg is conoidal, depressed at the top, flat at the base, shaped much like the eggs of Argynnis, but taller in proportion to the breadth, and the sides less rounded; marked by about twenty longitudinal ribs, half of which reach the summit, forming a serrated vein round the depression, marked by cross-striæ.

The young larvæ are cylindrical, thickest from joint 6 to joint 9. Color greenish yellow, each joint from 3 to 12 crossed transversely by two irregular rows of dark tuberculated spots or points on a pale ground, with a black hair from each. The second segment has a black stripe across the dorsum. Head black.

The color after moulting is reddish yellow, with two dorsal rows and one lateral row of indistinct whitish spots, which cover the junction of the segments and are in line with the spines. There are six rows of short, fleshy, tapering black spines, each beset with many short, fine black hairs. Collar reddish, edged with white. After the second moult the color is a shade darker and the white rows are more distinct. After the third moult the color is a little darker, shining, spines blue-black. Head bilobed, brown-black in front, red behind the vertices. It moults four times before reaching maturity.

The mature larva is 1.2 inches long, cylindrical. Color orange-ochre, smooth, striped longitudinally with black, which is almost concealed by the white spots that cover it. Two of these stripes are subdorsal, and another is just above the spiracles. Over the feet is a macular white stripe. Along the centre of the back, from joint 4 to joint 12, is a small white elongated spot, edged with black, over the centre of each joint. The spines are in six rows. The dorsals on joint 2 are orange at base, as are also those between the anterior joints; but all the rest arise from lustrous blue-black conical tubercles, and all the spines are blue-black thickly beset with fine, short black bristles. Between the anterior pair of dorsals is a black patch, and on the edge of the joint is a white spot. Head subcordate, front brown-black, vertices orange-red, with a patch of the same on the middle of the front.

The chrysalis is .8 of an inch long, of a pearl-white color, iridescent, marked with dark brown patches and points. On the abdomen there are four rows of conical tubercles.

There are two or three broods during a season, the last one probably hibernating in the larval state. It feeds on violet, passion-flower, mandrake, *Sedum, Desmodium*, and *Portulacca*.

New York to the Gulf of Mexico, Mississippi Valley, Colorado, Arizona, California; and occasional in New Hampshire.

47. MELITÆA PHAETON, Drury.

Expanse of wings from 2 to 2.25 inches.

Upper surface black, spotted with fulvous and pale yellow. The fulvous spots are a marginal row on both wings and in the discal cells. The marginal spots are nearly round on the fore wings, but on the hind wings they are blunt conical, the points inward. Those in the cells of the fore wings are two clusters, one of three at the end, and the other of two in the middle; and there is a cluster of four on the hind wings. There may be all of these present, or they may vary from this to none. There are also two or three more or less distinct spots on the costa of the hind wings. The yellow spots are two submarginal rows on both wings, the outer a series of crescents, the inner round; two half rows beyond the cell of the fore wings, and two or three spots in the cells. A part of these may be absent.

FIG. 44.

Melitæa Phaeton (natural size).

The under side is black, with the marginal row of orange or fulvous spots, two large spots in the cell of the fore wings, and six spots of the same in the basal half of the hind wings, arranged in two irregular rows.

Between these, or inside the marginal orange spots, are four rows of pale yellow spots.

Body black, a row of yellow spots on the side; palpi, club of antennæ, legs, and a broken stripe beneath on each side, orange.

The egg is obovate, truncated, rounded at base, flat or slightly depressed at the summit. The upper third is marked with from twelve to eighteen vertical ridges which gradually fade out below. Color lemon-yellow when first deposited, but changing in a few days to dull crimson, and finally, just before hatching, to black, this period being from nineteen to twenty days.

Fig. 45.

Egg of M. Phaeton, × 10.

The young larva is cylindrical, translucent, yellowish, a row of brown tubercles to each joint, from each of which arises a pencil of hairs.

The mature larva is from 1.1 to 1.3 inches long, cylindrical, the joints at the ends the smallest, the dorsum and sides armed with seven rows of long, tapering, fleshy spines, each of which springs from a round, shining, blue-black tubercle, the tubercles of each joint nearly meeting.

Fig. 46.

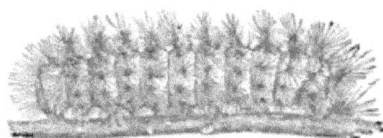

M. Phaeton, larva.

Each spine bristles with stout black hairs, giving the larva the appearance shown in Fig. 46. There is also another row of similar but much smaller spines below the stigmata; in this row joint 4 has no spine, joints 5 to 10 each have two in line, joint 11 has one, and joint 12 has a tubercle without a spine. On the under side of the body, on joints 5 and 6, in line

with the legs, is a single small tubercle with a short
branching spine. Joints 2 and 3, part
Fig. 47.   of joint 4, and the last two or three
joints are black; the others are deep
reddish fulvous striped transversely with
black.

The chrysalis (Fig. 47) is shaped
much as in Argynnis, the abdomen and
thorax furnished with several rows of
tubercles. Color white, marked and
spotted with brownish black, the tubercles orange.

M. Phaeton, pupa.

The food-plants are *Chelone glabra*,
Lonicera ciliata, and *Viburnum dentatum*.

United States east of the Rocky Mountains.

48. MELITÆA HARRISII, Scud.

Expanse of wings from 1.5 to 1.75 inches.

Upper surface of wings fulvous, the basal half and
terminal border black, with five fulvous spots in the
cell of the fore wings, two more below the cell, and
three in the cell of the hind wings. The base is not
wholly black, but is sprinkled with fulvous scales. The
border of the fore wings is broadest at the apex, where
it contains two pale dots; below this it extends inward
along the veins. Towards the anal angle it is more
broken up, so as to present a black edge and two indistinct lines.

On the under side the wings are fulvous, with a large
black subapical patch, which sends backward a subterminal band, with two rows of white spots extending
more or less through it. There are four black bars in

the cell and one beyond, and a white costal patch. Hind wings fulvous, with a median pale yellow band traversed by two black lines near the edges, so as to be three nearly complete bands. Inside this band are six spots of the same color,—two in the cell, three above, and one below. Beyond the median band is a subterminal row of crescents, with a row of black pupilled spots between the band and the crescents; all the light spots edged with black. Fringes white, black at the ends of the veins.

The eggs are lemon-yellow, in shape the frustum of a cone, with fifteen or sixteen longitudinal ribs which are elevated above the surface more in the middle than at either extremity.

The young larva is cylindrical, yellow-green, somewhat pilose; head obovoid, bilobed, the vertices rounded, dark brown.

After the first moult the larva is armed with seven rows of short black spines, tapering, and thickly set with short black bristles. Color yellow-brown. Very little change takes place after the second moult, save that the color is ochre-yellow, with five transverse black stripes on each segment. After the fourth moult the color is red or orange ochraceous, striped as before, three to a segment. In coming to maturity it moults five times.

The mature larva is of a deep red fulvous color, crossed by black stripes, one before and two after each transverse row of spines, and with a dorsal black stripe. The last two joints are nearly all black, and on joints 9 to 11 the fulvous bands are spotted. The spines are in seven principal rows, with a row of smaller ones just above the feet. The spines are long,

Fig. 48.

M. Harrisii, larva.

tapering, black, each thickly set with long, divergent black hairs.

FIG. 49.   The chrysalis (Fig. 49) is cylindrical, but with a small depression on the back of the thorax, abdomen with several rows of sub-conic tubercles. Color pure white, marked and spotted with black, or brown-black and orange.

M. Harrisii, pupa.   The food-plants are Aster and *Diplopappus umbellatus*, and the imago is to be seen in June.

New England, New York, Michigan, Wisconsin, Illinois.

49. PHYCIODES NYCTEIS, Doub.—Hew.

Expanse of wings from 1.5 to 1.75 inches.

Upper surface fulvous, a broad black terminal border, on the fore wings broadest at the apex, enclosing a sub-

FIG. 50.

Phyciodes Nycteis, male (natural size).

terminal row of fulvous spots, more distinct in the female than in the male. At the end of the cell is a broad black patch connected by a line with a smaller one on the sub-median vein (see Fig. 50). In the cell and below it several indistinct black marks, the base black, this extending outward along the costa and hind margin to the black patches.

The hind wings have the basal half marked the same as the fore wings, though sometimes more suffused.

There is a broad terminal border almost meeting the basal black on the costa, and sending a shade across the wing through the fulvous space, also a subterminal row of black spots, some of which are pupilled.

Under side of fore wings pale fulvous, with three or four not very prominent bars in the cell, a somewhat triangular patch beyond, and a terminal brown-black border with the subterminal row of spots, three apical silver-white lunules, and two more marginal near the middle, the lunules resting on a terminal yellow line edged within with brown.

The hind wings are dark brown broken by pale yellow, especially in the basal portion, where it is the principal color. A row of large silvery white spots crosses the wing near the base, and a similar band crosses the middle of the wing, broken by brown veins and edged on the outside by a crenate brown line, and a marking of pale yellow beyond. The subterminal row of round black spots is reproduced, part pupilled with white. There is the terminal yellow line the same as on the fore wings, with a row of silver-white lunules, the middle and two apical much the largest.

The eggs are deposited in clusters of about a hundred on the under side of the food-plant. They are whitish green, somewhat in the form of a truncated cone, the lower third of the outside smooth, the middle part marked with hexagonal cells, and the top by longitudinal ribs. They hatch in from nine to thirteen days.

The young larva is .06 of an inch long, with a dark brown head, and a yellowish green body clouded with brown, with scattered black hairs. After the first moult it is smoky brown, and, like Melitæa, armed with seven rows of stout, fleshy, tapering black spines, and a minute row over the feet. After three or four days it moults a second time, when the color is black-brown, and this color remains through the next stage, with sometimes a broken yellow stripe along the side. To come to maturity it moults four times.

The mature larva is an inch in length, blackish brown above and greenish brown beneath. Head black, cordate, the sides high and rounded, and clothed with numerous black hairs arising from black papillæ. The spines long, black, tapering, armed with short black hairs, each springing from a shining black tubercle, except those of the lowest row, which stand upon greenish or yellowish tubercles. A black band runs along the base, with a yellow stripe in the line of the lower lateral spines, and a broken yellow stigmatal stripe. In some cases this is ochre or reddish yellow. The back and sides are much dotted with white.

The chrysalis is similar in shape to *M. Phaeton*, with five rows of conical tubercles on the abdomen. The color varies extremely : some are wholly greenish yellow, others pink-brown, others gray-brown ; with usually but few dark markings.

The food-plants are *Diplopappus umbellatus*, Aster, *Actinomeris*, and sunflowers.

Maine to North Carolina and west, Mississippi Valley.

50. PHYCIODES CARLOTA, Reak.

Expanse of wings from 1.3 to 1.5 inches.

Upper surface much as in *P. Nycteis*, except that most of the wing is black, there being some fulvous spots near the base of the fore wings, a fulvous band through the middle crossed by the black veins, a subterminal row of whitish dots in the broad black terminal border, and a white lunule in the middle of the border near the edge of the wing. In the female there are traces of other white lunules just within the margin. The hind wings are similarly marked, except that there is a row of black submarginal spots circled with fulvous, and the lunules are faint in the males.

Under side of fore wings fulvous, marked about the same as in *P. Nycteis*, except that there is more black through the middle. The terminal border is as above, save the terminal lunules. Between median venules one and two there is a large conical whitish spot, the base on the double terminal line; this double line running in zigzag to the apex, the inner points silvered more broadly towards the apex, and the inner point sending a white ray to the margin. The same is repeated towards the posterior angle, but with less silver.

The hind wings dark brown washed with whitish, more towards the base, only two yellowish spots in the cell. Near the base is a broken silvery band, and through the middle a silver band crossed by the brown veins, the outer margin dentate. The submarginal row of black spots as above, only they are pupilled with white and not circled with fulvous. The margin is similar to that of the fore wings, save that the large spot is silvery instead of whitish.

Southern and Western States, Rocky Mountains, Montana to Arizona; occasional in West Virginia.

51. Phyciodes Phaon, Edw.

Expanse of wings from 1 to 1.4 inches.

Upper surface of fore wings fulvous, the base, costa, hind margin, outer margin, and a band through the middle black. The basal half of wing contains several irregular black marks, and the median black band is expanded on the costa and hind margin. Beyond the median band there are two light bands crossed by the black veins and separated by a black shade which expands into a triangle on the costa and hind margin. The first light band is pale fulvous, almost buff; the second is the same fulvous as the ground color, and contains a black dot near the posterior angle. The outer edge brownish black, with a pale lunule in the middle of the border.

Hind wings similar to the fore wings, only the inner margin is fulvous, the median black band is narrower, both light bands are fulvous, and the outer contains a row of black dots. In most specimens there are only a few narrow whitish lunules near the anal angle, but sometimes these continue to the apex.

The under side of fore wings is orange fulvous, the basal half marked with a darker shade, the median black band as above, and also the two light bands, only the inner is more whitish and the shade separating them is obliterate except the triangles. Terminal border black, pale at the edge, with a crenate black line near the edge, and a pale yellow patch in the middle, and one at each end.

There are two forms of this having the under side of hind wings differently colored. The winter form is pale buff washed with umber-brown, the basal half with more or less complete bands of pale spots edged with brown; the row of black spots the same as above; a submarginal row of lunules, the middle large, silvery, the others more or less obscure. The summer form is pale buff, with irregular transverse brown lines, the brown dots smaller, a terminal brown border accompanying the submarginal row of lunules and partly obscuring three of them, a small brown patch on the costa and sometimes a little in the centre.

Gulf States, Texas; occasional in Kansas.

52. Phyciodes Tharos, Drury.

Expanse of wings from 1.15 to 1.5 inches.

There are two dimorphic forms of this species, the winter form, *Marcia*, and the summer form, *Morpheus*. It was supposed that these two forms were distinct species till Mr. W. H. Edwards proved by rearing them that they are seasonal forms of one species, the difference in coloration being due to the effects of cold while hibernating. Besides these two well-marked forms there are several minor variations, only one of which is named.

Winter form, Marcia, Edw.—This has the upper surface reddish fulvous marked with black. There are two rows of more or less distinct coalescing circles near the base of the wing, the first of two circles, the second of four, and an ellipsoid at the end of the cell; a patch of black beyond the cell on the costa and one on the hind margin, sometimes the two being connected by a dentate line. The outer border is broad, black, and through it runs a crenated line with a yellowish or fulvous lunule in the middle, in some examples distinct, in others connected with the central color; a black dot near

m

the posterior angle. Hind wings very much as in *P. Phaon.*

The under side of fore wings is yellowish fulvous, with yellow spots and four black patches,—two on the costa before the apex, one at the posterior angle, and one on the hind margin; the outer margin with the lunules as in *P. Phaon,* only yellow.

The hind wings resemble closely those of the winter form of *Phaon,* the yellow perhaps a little deeper. The terminal, costal, and middle brown patches are present in some examples, the wing being well suffused with brown.

Summer form, MORPHEUS, Fab.—This is scarcely distinguishable from *Marcia* on the upper side, though the black is more inclined to be in lines. The under side of fore wings like *Marcia.* Under side of hind wings yellow-buff, the brown patches on the costa and in the cell absent, though in some specimens there is a slight discoloration at the end of the cell. In some females a slight costal patch is present.

Aberr. PACKARDII, Saund.—This differs from the usual forms in the wings being brown above, with a cupreous tinge and sprinkled with fulvous atoms. The fulvous is in bands: first a macular band near the base of the fore wings, not quite reaching either margin; a patch across the outer part of the cell; and a wide band beyond the cell, crossed by black veins, narrow on the costal end. On the hind wings a large fulvous patch covers about the inner half, containing several roundish black spots; beyond this is a macular band of fulvous between two broad brown bands. Under side pale.

The eggs are deposited in clusters of two hundred or more on different species of Aster, both wild and cultivated. They are pale green, conoidal, depressed at the top and rounded at the base. The lower half is indented like the surface of a thimble, the upper half has about fifteen ribs. They hatch in from four to seven days.

The young larva is yellowish green clouded with brown, with a dark brown head. It is covered with scattered black hairs.

After the first moult the larva is armed with seven rows of short, fleshy, brown spines, each thickly set with short concolored bristles; also at the base of body a row of small spines. Body striped longitudinally with light and dark brown and sordid white. Head subcordate, the vertices rounded; two gray bands, the rest black. There are but few changes after the second and third moults.

The mature larva is .85 of an inch long, with a cordate, shining, bronze-colored head, having two oblique white stripes on each side and a spot of the same color above the mouth. The body is dark brown dotted with yellow, and has seven rows of tapering fleshy spines armed with blackish bristles.

The first brood passes four moults before reaching maturity, when it changes to a chrysalis, from which the imago emerges in from seven to thirty days. The larvæ of the second brood pass three moults, when they become lethargic and hibernate. In the spring they revive, go to feeding, and moult twice more before reaching maturity, the chrysalides from these producing the butterflies in from one to two weeks.

The chrysalis is about half an inch long, cylindrical, with a deep depression back of the mesonotum, and several rows of fine tubercles on the abdomen. The color varies much, being light cinereous covered with fine abbreviated brown streaks; or cinereous on dorsum, the abdomen and wing-cases tinted with yellow-brown; or dull white clouded with brown; or wholly dark brown speckled with gray.

The butterflies are to be seen from May through the season.

United States generally, except the Pacific States.

53. PHYCIODES BATESII, Reak.

Expanse of wings from 1.25 to 1.63 inches.

Male.—Fore wings black, two transverse maculate fulvous bands beyond the cell, the inner pale, arising nearly at the costa and converging to the hind margin. The cell contains three or four abbreviated bars, none of them extending below the median vein. The basal half of the area below this vein is deep black, rarely containing a narrow fulvous streak; a spot of fulvous in the middle of the outer margin.

Hind wings black, the two fulvous bands of the fore wings continued nearly to the inner margin, much wider than on the fore wings, the separating line very much attenuated in the middle. The outer band contains a series of rounded black spots between the venules, and beyond these an indistinct row of connected lunules. There are two fulvous spots within the cell, preceding the first transverse band, the inner semicircled by the outer. Fringes white or whitish, more or less cut with black at the ends of the veins.

Under side of fore wings fulvous; a large triangular black patch upon the middle of the hind margin is connected at its apex with an irregular, dilating bar running thence to the costa; a short bar between this and the apex. Three connected black lunules, the central much the widest, run from beyond the middle of the hind margin to the third median venule. In some this line is prolonged by the addition of one or two more very delicate crescents.

Hind wings ochrey yellow, with indistinct pale fulvous lines near the base, and a row of rounded dots followed by pale lunules of the same color near the outer margin; rarely one of the last is bright ferruginous.

Female.—Similar to the male on the upper surface, the inner band of the fore wings paler than in the male. On the under side the reticulations are plainer.

West Virginia, Maryland, New York to Ohio.

54. ERESIA FRISIA, Poey.

Expanse of wings 1.4 inches.

Upper surface reddish fulvous, the base dusky. Across the inner third are four more or less distinct sinuous black lines, which are fine and nearly distinct on the hind wings, but are heavier and more blended on the fore wings, and in places connected by cross-lines. Beyond the basal third the fore wings are crossed by three black bands and a terminal border, the first and second united into a broad band at the end of the cell, reaching from the median vein to the costa, with a fulvous spot at the end of the cell just within the edge of the band, the two bands approaching each other near the submedian; the third band expanded from the costa back to the fifth subcostal venule, from which it gradually contracts across two interspaces, expanding abruptly at this point, where it unites with the second band, to separate again below the second branch of the median, from which it continues without much variation to the hind margin. The spaces between these bands are paler fulvous than the base and the hind wings, the subterminal being whitish. The first and second of these bands of black are continued across the hind wings in black shades hardly positively enough marked to be called lines. The broad terminal border of the hind wings contains a series of connected whitish lunules.

The under side has a little more than the basal half of the fore wings fulvous, with four or five large whitish spots; the terminal portion dark brown, with the border whitish and two large whitish patches. The hind wings are marked much like a Phyciodes, an oblong brown shading from the base along the middle of the wing to the centre; a whitish band marks the outer third, a sub-terminal row of whitish lunules, before which is a series of brown sagittate spots.

This resembles to some extent some of the forms of *Phyciodes Tharos*, but the difference may be seen by comparing the descriptions.

Found at Key West, Florida, and Cuba.

55. GRAPTA INTERROGATIONIS, Fab.

Expanse of wings from 2.3 to 2.75 inches.

This is a dimorphic species, the hibernating form being known as form *Fabricii*, the other as *Umbrosa*. There are about four broods in a season; and while the last brood or hibernating butterflies are the pale forms, the others are more or less mixed, as Mr. Edwards has shown.

Dimorphic form, FABRICII, Edw.—This has the upper side fulvous, marked with ferruginous brown and spotted with black. The fore wings have two black spots in the cell, one round, the other a short bar, and a wide bar at the end, broadest at the costa; and a row of four spots from the hind margin to beyond the cell, all but the last round. Outer border ferruginous brown, with a subapical bar of the same.

Hind wings with four more or less distinct median black spots, and a broad ferruginous brown border enclosing a submarginal row of fulvous spots. Edges of both wings whitish purple.

The costa is rather falcate, the apex truncate, and the hind wings have a short tail.

Under side clouded in shades of brown, in some examples partly suffused with purple, without the striking diversity of color found on *Umbrosa*, the common row of black points more or less obsolete, costal edge of fore wings near the base clouded with small yellow spots, with some yellow color below. Discal silvery or golden mark on the hind wing an interrupted C.

Dimorphic form, UMBROSA, Lintn.—This has the fore wings as in *Fabricii*, but with less purple edging. The hind wings have the outer two thirds overlaid with black, in some examples quite intense, the tail purple.

The under side is variegated with ferruginous brown, olivaceous, and more or less purple. The basal third is limited by an irregular ferruginous, partly olivaceous band, within which the ground color is yellow-brown streaked with ferruginous. Beyond this band the fore wings have a narrow belt of yellow-brown, and beyond this the colors are brown and olivaceous. There is a lilac patch near the posterior angle, and another sub-

apical. The costal margin of the hind wings is olivaceous, with a median band similar to that of the fore
wings; the inner margin and tail portion of the outer
margin lilac. Crossing both wings is a row of black
points, those on the hind wings within a brown ferruginous band.

The female differs from the above in having the under
side more of a brown suffused with blue-gray.

The eggs are pale green, conoidal in form, with the
base flattened. The sides are rounded, and marked by
eight or nine ribs, which are low near the base, but higher
above, terminating abruptly around a small flat space at
the top.

The young larvæ are whitish yellow, somewhat marked
with brown, head shining black. After the first moult
their color is black more or less specked with white, and
they begin to be clothed with short spines, all black except
those on the eighth and tenth segments, which are whitish.
After the second moult they begin to assume the type
they retain to maturity. The spines are in seven rows,
fleshy at base, slender and many-branching at extremity;
the dorsal and first lateral on joint 3 are black, on joints
2, 4, and 11 russet, the rest yellow; the second laterals
black throughout, the lowest row greenish; head bilobed,
black, with short black spines at vertices. After the third
moult the larvæ vary greatly both in color of body and
spines. Some are black finely specked with yellowish;
others are yellow-brown specked with yellow tubercles;
others gray-brown, with indistinct reddish lines between
the spines on the dorsal and two lateral rows, and much
tuberculated; others are black, with fulvous stripes and
profusely covered with yellowish tuberculated spots and

points. The spines vary from black to fulvous and
green and yellow. After the fourth moult
the larva feeds four or five days and changes
to a chrysalis.

Fig. 51.

The chrysalis (Fig. 51) is an inch long,
variable in color from light yellowish to
dark brown; the head deeply notched, a
thin prominence on the thorax, and eight
silvery spots on the back.

The food-plants are hop, elm, nettle, false
nettle, and basswood.

G. Interroga-
tionis, pupa.

United States generally, except the Pacific
States; Arizona.

56. GRAPTA COMMA, Harris.

Expanse of wings from 2.25 to 2.5 inches.

Upper surface fulvous, the outer border black, with a
little edging of lilac. The fore wings have a subapical
patch of brown, and another at the posterior angle, each
enclosing a fulvous spot. Like *G. Interrogationis*, this
species has two spots in the cell and a bar at the end,
but there are only three round spots between the cell and
the posterior angle, the lowest supplemented by a shade
above it.

Hind wings with ferruginous brown next the border,
shading out towards the middle, with a series of pale
fulvous spots next the border, and an irregular row of
black spots across the middle.

The under side is marbled with light and dark brown,
and washed with olive, and in the males with more or
less of pink. Across the middle the shades are darker,
clearly defined externally, beyond which it is washed with

16*

pink. The hind wings have a discal silvery C. Both
wings have the submarginal row of dots in a more or
less distinct band of olive and ferruginous brown. In
the female these shades are not so distinct, with less
pink, in some specimens the whole surface being washed
somewhat with blue-gray.

This is a dimorphic species, the last brood of the
summer, or the one that hibernates, being known as the
winter form, HARRISII, Edw., from which the above
description is taken. The summer forms are called
DRYAS, Edw., and differ from the others in having the
hind wings above suffused with black, as in *Umbrosa* of
the preceding species, and the under side more suffused
with brown.

This species feeds on the same plants as *G. Interrogationis.*

The eggs are green, and similar in form to those of
that species, with ten longitudinal ribs and cross-striæ
between them.

The young larvæ are one-tenth of an inch long,
black, covered with short hairs. After the first moult
the color is either brown-black or black, with whitish
lines at the junctures of the segments; armed with seven
rows of branching spines, stout, black, and beset with
short bristles. In the black examples all the tubercles
are black; in the specimens with white lines, on segments 4, 6, 8, and 10 the spines spring from whitish tubercles. Besides these there is a row of minute spines
over the feet. Head dark brown. After the second
moult the color is dark olive-brown, or black-brown, or
reddish brown, two or three fine white transverse lines
to each joint, and two white bars on the back. After the

next moult the color is black, with the stripes the same, and a yellow substigmatal band.

The mature larvæ are from an inch to an inch and a quarter long, and quite variable in color. Some are black, with yellow bases to the spines, others are nearly white, with red spots along the sides, while still others have a reddish or vinous tint instead of black.

The chrysalis is about four-fifths of an inch long, of various shades of gray or brown, with golden protuberances on the abdomen, and a flattened prominence on the head.

Eastern, Middle, and Northwestern States; North Carolina, Tennessee, Kansas to Texas.

57. Grapta Faunus, Edw.

Expanse of wings from 2 to 2.25 inches.

Upper surface fulvous. Next the apex of the fore wings, the base of both wings, and the inner margin of hind wings dusky. The fore wings have a broad black outer border, dentate at the apex, and bordered within by a series of subobsolete tawny lunules. The markings on the wings are much as in *G. Comma,* but are heavier and more black and less brown.

Under side of both wings dark brown on the base, with an irregular common blackish band across the middle, darkest on its outer edge and within the abdominal margin, where its outline is obliquely serrated. Beyond this band the color is pale brown mottled with grayish white, which is clearest on the fore wings. The whole surface is clouded with vinous, and more or less crossed by fine abbreviated streaks of brown. Apex of fore wings yellow-brown, with three small lanceolate ferruginous spots, the lower enclosing a blue or green point. The outer margin of both wings, below these, is bordered by a series of confluent blue-black—sometimes olive-green—spots following the outline of the wing. Within these is another series of rounded spots of the same color. On the disk of hind wings is a white G varying in form.

The mature larva is one inch and a quarter long; head black, furnished with two branching horns and a few scattered white hairs. Upper side of joints 2 to 6 brick-red striped transversely with blue, yellow, and black, a few white hairs on joint 2. There are four branching yellow spines, with black tips, on joints 3 and 4, and six on joints 5 and 6. Joints 7 to 12 are white, with a faintly-marked black stripe along the back, each joint with three transverse yellow bands and two oblique black spots. These joints have each seven spines, all white except the one next the lowest, which is brown. Last two joints black, with seven and four spines respectively. Sides red, with two black bands, the lowest spotted with blue.

The chrysalis (Fig. 52) is grayish brown. Head with two biforked horns, the outer point very short; thorax with an elevated keel-like ridge on top, with a small tubercle on each side. At the base, below this, there is a larger tubercle, and behind it another keel-like protuberance, hollowed on top; there are six raised silver ornaments on the dorsal surface, the first resembling in shape a capital G; the second is an oblong spot, and the third is a sharply-pointed tubercle. The abdominal segments are furnished with eight rows of tubercles; on each side are five

FIG. 52.

G. Faunus, pupa.

brown spots, and below the spiracles there is a brown stripe.

The food-plants are gooseberry, currant, willow, and birch.

Mountains of New England and New York, Michigan, Nebraska, Washington Territory; occasional from West Virginia to Georgia.

58. Grapta Gracilis, Gr.—Rob.

Expanse of wings 2.25 inches.

Upper surface fulvous, darker at the base and fading out paler in the outer half, the fulvous brown border of the fore wings replaced on the hind wings by yellowish edged internally by ferruginous brown. Next this border is a series of elongate yellow lunules, confluent on the hind wings, where they are supplemented internally by a rusty brown shade. On the fore wings there is the usual subapical brown patch, also one at the posterior angle, usually connected with a shade of the same. In the middle area of the fore wings are the usual five dots and bar of black, and three black spots on the hind wings.

The color of the greater part of the under side is olivaceous yellow, with the usual vinous brown shade through the middle, sharply defined outwardly, beyond which the wing is paler. There are three elongate patches of this color, edged with darker, in the cell of the fore wings, and the base of both wings is marked with the same. The whole wing is marbled with fine abbreviated streaks of either brown or a darker shade of the ground color. There is the usual submarginal row of points in an olivaceous band, the three subapical

preceded by white shading. Next the margin, along
the middle of both wings, is a series of elongate lunules,
dark olive or greenish, the posterior and anal angles
washed with white. There is also the usual discal G to
the hind wings.

This is found in the White Mountains of New Hampshire.

59. GRAPTA PROGNE, Cram.

Expanse of wings from 1.75 to 2 inches.

Upper surface bright fulvous, a little paler near the
extremities. The border to the fore wings is black or

FIG. 53.

Grapta Progne,—right wing the under surface.

blackish brown, brown at the apex, with the yellow
lunules and subapical and posterior patches as in *G.
Gracilis*, as also the black spots in the central area, as
shown in Fig. 53. The outer half of the hind wings
blackish, shading with the fulvous; the yellow lunules
of *Gracilis* showing more or less as points, with the
black central spots more or less distinct.

Under side grayish brown, closely streaked with fine
abbreviated lines of dark brown, with the usual median

dark brown shade. Beyond this on the fore wings the
ground color is pale gray, whitish near the costa, with
the usual row of points. Just within the edge is a row
of confluent crescents, greenish edged with black within.
Discal G slender and somewhat abbreviated.

The egg is conical, the base rounded; marked by eight
or nine vertical ribs, which gradually increase in prominence from the base upward and are crossed by many
transverse striæ.

The young larva is at first dull green, the last joint
with a brownish tint, but later it becomes dull white and
brown, with the usual black tubercles and black cervical
spot. After the first moult the color is greenish brown,
with seven rows of large branching spines. All are
black, but they arise from light yellow tubercles, and are
yellow about half-way up; on joints 12 and 13 they are
almost wholly yellow. Head bilobed, black. After the
second moult the color is at first yellow, but in twelve
hours changes to brown with white cross-lines. After the
third moult it is glossy black from joints 3 to 11, crossed
on the posterior half of most of the joints by three white
lines, with white or gray oblique stripes on the ridges on
which the spines stand from joints 5 to 11.

The mature larva is from 1 to 1.2 inches long, of a
buff color, the cross-stripes on the posterior parts of the
joints black and pale buff. In front of each dorsal
spine is a V-shaped reddish bar with the spine within
the angle, and an oblique bar of the same color in front
of each of the laterals, from its base, directed forward
and downward. The second laterals stand on a straight
or slightly-arched bar of the same color. The spines on
joints 3, 4, and 5 are the largest. The dorsals are white,

yellow at the bottom,—the yellow being reddish or honey-yellow,—and arise from yellow tubercles. The first laterals are white from joints 5 to 11, but those on joints 3, 4, and 12 are black, with buff branches. The second laterals are black, with yellow bases and tubercles. The spines on joint 2 are yellow. Head subcordate, on each vertex a large compound spinous process, the body black, the branches partly black and partly yellow.

The chrysalis is similar to that of the other species, and is .7 of an inch long, with several rows of tubercles on the abdomen. Colors dull green, brown, and pinkish white. There is much variation in color of both larvæ and pupæ.

The food-plants are gooseberries and currants, and there are two broods in a season, the last brood of butterflies hibernating.

Northern and Western States.

60. GRAPTA J ALBUM, Bd.—Lec.

Expanse of wings 2.75 inches.

Upper surface dull yellowish, dusky at base, washed more or less with rusty brown, especially on the basal half. Outer border of fore wings dull black, with a double crenate line, and two more or less obscured large costal bars,—one at the end of the cell, and the other between the cell and a white subapical patch; a black spot in the cell, with three black spots below the cell, the one nearest the base of the wing quite large.

The hind wings have a black and brown border a little within the margin, the yellow outside sprinkled with brown atoms; a large black space below the costa, beyond which is a whitish patch.

Under side grayish brown, the usual darker band across the middle, which on the hind wings is but little darker than the base; beyond this greenish white, the whole surface finely reticulated with brown lines. There are the usual three elongate patches edged with dark brown in the cell of the fore wings, the submarginal row of ashy-blue lunules edged with dark, and the row of points between these and the median band. The lunules do not reach the apex of the fore wings.

The mature larva is two inches long, head with black markings on the sides, thickly set with little points and with short spines; somewhat cordate, the vertices surmounted by two shining black, thick spines, verticellated near the tip. The dorsal and subdorsal rows of spines shining black except at the base, which is reddish, with long branches, those of the anterior joints more thickly branched than the others. The super- and sub-stigmatal rows reddish tipped with black.

The chrysalis is one inch long, of a beautiful green color, delicately reticulated, with six golden spots on the back. The spines and projections are similar to those in *G. Comma*.

The food-plant is not known, but Professor Fernald's surmise is that it may be willow. The butterfly hibernates, the new brood appearing about the middle of August.

Northern States to Wisconsin.

61. VANESSA ANTIOPA, Linn.

Expanse of wings from 2.5 to 3.5 inches.

Upper surface rich dark maroon-brown, the border yellow sprinkled with brown, and preceded by a black band containing a row of blue spots. The costa is mottled with yellow and contains two yellow patches.

Under surface traversed by numerous fine black abbreviated lines, the outer margin pale buff sprinkled with brown, and preceded by a series of confluent gray, blue-black-edged lunules.

Aberrant form, LINTNERII, Fitch.—This form differs from the one usually seen in having the outer pale border twice as wide as in the typical *Antiopa*, occupying one-third the length of the wing, and in being wholly destitute of the blue spots. The general color is more reddish ; the costal margin is black, with small whitish transverse streaks, but destitute of the two patches.

Another form has "the broad outer border of a tarnished pale ochre-yellow hue, speckled with black the same as *Antiopa*, and becomes quite narrow at the anal angle. The wings beneath are similar to those of *Antiopa*, but are darker and without any sprinkling of ash-gray scales or any whitish crescent in the middle of the hind pair, and the border is sprinkled with gray whitish in wavy streaks, without forming the distinct band which is seen in *Antiopa*." This is Mr. Bunker's description, stripped of a few superfluous words, of a specimen taken near Rochester, New York.

Fig. 54.

Cluster of eggs of V. Antiopa, × 2.

The female deposits the eggs in a cluster round the twigs of willow, elm, or poplar near the petiole of a young leaf, upon which the young larvæ may feed. The mature larvæ are two inches long, black, minutely dotted with white, which gives them a grayish look ; with a dorsal row of brick-red spots. Head black, roughened with small black tubercles. The spines on the body are black, rather long, slightly branching. There are four on joints 2 and 3, six on joints 4 and 5, and seven each on joints 6 to 12. The last joint has two pairs of short spines, one behind the other.

The chrysalis is dark brown or gray, with two rows of spines along the back of the abdomen, two on the head in front, three on the edge of the wing-covers on each side, and a thin prominence on the middle of the thorax.

This species hibernates in the butterfly state, and the first brood of caterpillars may be seen in June. The second brood of caterpillars appears in August.

North America generally.

62. VANESSA MILBERTII, Godt.

Expanse of wings from 1.6 to 2.25 inches.

Upper surface brownish black, with a broad fulvous band between the middle and the outer margin, paler on its inner edge. One example from Colorado has fully half the band pale buff. On the fore wings the pale band contains a black patch on the costa, with a white spot on one or both sides. There are two fulvous spots in the cell. The border is composed of two parts, the inner black, the outer a black-brown crenate line, on each side of which it is a little paler. The black on the hind wings supports a row of violet lunules.

Under side dark brown, with the usual wavy lines and spots; the outer half yellowish brown, differing in shade on different specimens, with a submarginal row of gray-blue lunules which are black-edged.

The mature larva is a little more than an inch long, with a black head sprinkled with minute whitish dots, from which spring pale hairs. The body is nearly black above, with small white dots and pale hairs, which give it a grayish color. The spines are arranged as in *V. Antiopa*, and are black and branching. It has a greenish-yellow lateral line, above which is a broken line of brighter orange-yellow shade.

The chrysalis is .8 of an inch long, slightly angular; the frontal beaks short, conical; thoracic projection forming nearly a right angle; dorsal spines but little elevated,—the superior one exceeding very little the others in size; wing-cases as in *V. Antiopa;* terminal spine short, flattened, curved.

The larvæ are to be found on the wild nettle, and there are two broods in a season.

Northern States to Montana, Colorado, Arizona, Pacific States.

63. PYRAMEIS ATALANTA, Linn.

Expanse of wings from 2.25 to 3 inches.

Upper surface black, a little brownish over the basal half. Each wing is crossed by a reddish fulvous band, the one on the fore wing extending in a curved line from the costa at one-third the distance from the base of the wing almost to the posterior angle; the one on the hind wing a terminal border not reaching the apex nor the anal angle, and containing a row of black lunules. The fore wings have an oblique white costal bar half-way from the fulvous band to the apex, and a submarginal row of fine round white spots from the costa to near the fulvous band, the fourth the largest. Near the anal angle is a blue bar in a black space.

Under side of fore wings black, gray at the apex, the

fulvous and white repeated, blue along the costa, in the cell, and beyond the fulvous band. The hind wings are marbled with brown, olive, olive-brown, gray, and pale violet, a series of five partially distinct submarginal ocelli imperfectly pupilled.

The eggs are green, barrel-shaped, with nine vertical ribs which are highest at the top, the ribs grooved on each side perpendicularly to the surface of the egg.

FIG. 55.

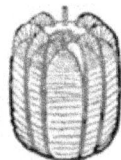

P. Atalanta, egg, × 20.

The young larva is greenish brown, semi-translucent, and furnished with ten rows of black, curved hairs. Joints 2 and 13 have black dorsal patches.

After the first moult it is wholly black-brown, armed with seven rows of short, slender, branching black spines. Head bilobed, the vertices rounded. After the third moult the body is more black, each segment creased, on the creases many minute whitish tubercles; a macular greenish-yellow stigmatal band; head brown. In reaching maturity it passes four moults.

The mature larva is 1.3 inches long, cylindrical, enlarged in the middle, and of a velvet-black color thickly sprinkled with fine yellow points, with a stigmatal line of greenish-yellow patches. It has seven rows of moderately long, slender, branching spines, which are usually black; but in some specimens they are pale yellow-white, more or less reddish at base. Head rounded, bilobed, the vertices rounded, thickly covered with black, simple spines.

The chrysalis is from .85 to .95 of an inch long, cylindrical, the abdomen stout, the dorsal tubercles gilded, the lateral in two rows and black. Color various, usually reddish gray, more or less reticulated with black.

The food-plants are nettle and hop.

United States generally.

64. Pyrameis Huntera, Fab.

Expanse of wings from 2 to 2.25 inches.

Upper surface fulvous, a little tawny at base, the apical portion of fore wings black, this continued as a border to the posterior angle. The apical black contains an oblique fulvous bar beyond the cell, and the submarginal white dots of *P. Atalanta*, the first two blended, and one farther down in the fulvous. Besides this there are five triangular black marks, two of which are in the cell. The border of both wings consists of three more or less distinct lines, the inner on the hind wings in the form of a shade. The apical portion of the border on the fore wings is washed with lilac; and there is a gray-blue bar at the anal angle. Hind wings have a submarginal row of five black spots, the second and fifth pupilled with blue. Costa black.

The under side of fore wings is red, except the apical portion, which is marked as above. Hind wings marbled with brownish black and white, with two large ocelli. The outer border is four black lines, with violet between the two inner.

The mature larva is 1.25 inches long, the body velvety black, between the joints four transverse lines of pale yellow alternating with narrow black lines. On each joint there are seven dark brown tubercles, from which arise short, branching, black spines. On joints 6 to 12 in the subdorsal region are large shining white spots.

Joint 2 is short, has but little black, and lacks the tubercles and spines. Head bilobed, black, sparsely clothed with gray hairs. Between joints 12 and 13 is a large whitish patch crossed by a fine black line. Below the stigmata are two yellow lines, the lower interrupted, both spotted with black.

The chrysalis is yellowish, moderately angular; head-case bifid, slightly projecting, and edged with brown above; thoracic projection forming an obtuse angle; dorsal spines minute, of nearly uniform size, brown-tipped; segments with rows of brown dots, and also of brownish markings.

The food-plants are *Gnaphalium polycephalum*, *Artemisia Ludoviciana*, and probably other allied species.

United States generally.

65. PYRAMEIS CARDUI, Linn.

Expanse of wings from 1.75 to 2.5 inches.

Upper surface reddish fulvous, the fore wings marked as in *P. Huntera*, except that the bar in the apical black portion of the fore wings is white instead of fulvous, and the white submarginal dot in the fulvous is absent, as is also the violet apical shade. The hind wings have the submarginal black spots, with a very little blue in the fourth and fifth, and the border is broken.

The under side is much as it is in *P. Huntera*, but there are five ocelli on the hind wings instead of two, and they are smaller.

The mature larva is 1.5 inches long, cylindrical, rather robust. The general color of the substigmatal region, as well as that of the middle part of each joint, and the greater part of the thoracic joints, is a delicate lilac. Between the joints are two lines of bright yellow, the posterior slightly double, the space between the yellow lines occupied by two narrow black lines and one white line, all the light lines between the thoracic joints being white. There are seven rows of tubercles, from which arise branching spines, the dorsal and lowest lateral tubercles white, the rest reddish brown. The spines are whitish yellow, the tips of the branches black. Joints 3 and 4 have only four spines each. The dorsum has a double, broken yellow line, the dashes of which it is composed extending from the anterior transverse yellow line to the tubercle on the centre of the joint. There is also a yellow dash in front of each of the brown tubercles; all the yellow being rather dark. Below the stigmata, between the lower tubercles, is a light lemon-yellow line. Stigmata black, with some black spots over the body. Head black.

This butterfly is distributed over the United States generally, and is known by the common name of Thistle Butterfly. It is double-brooded, and hibernates in the butterfly state. The larvæ feed on thistle, burdock, sunflower, and hollyhock.

66. JUNONIA CŒNIA, Hüb.

Expanse of wings from 2 to 2.5 inches.

Upper surface dark olive-brown, each wing with two eye-spots, a large and a small one, the large ones the posterior of the fore wings and the anterior of the hind wings, the small ones on the fore wings sometimes obscure. There is an oblique whitish band beyond the cell of the fore wings, the lower part expanding so as more or less to enclose the eye-spot. There are two fulvous bars in the cell, and there is a little fulvous outside the large eye-spot, and a subterminal fulvous border to the hind wings not reaching either margin. The border to both wings consists of three somewhat crenate black lines, the ground color between a little pale. Sometimes the fore wings have a little subapical white, and a subterminal fulvous bar outside the small ocellus.

Under side variable, ranging from reddish brown and brownish fawn to brownish buff; these colors being found on the hind wings and the apex of the fore wings, with

FIG. 56.

Junonia Cœnia (natural size).

brown wavy lines of varying shade. The eye-spots of the fore wings are as above, but both of those on the hind wings are small, with two points between them and one near the costa. The fore wings have the white band and three fulvous bars in the cell.

The mature larva is black, the dorsum sprinkled with fine white specks, with two somewhat broken lines of creamy white on each side, the lower side of the joints back of joint 4 and a ring back of the head dull ochre.

On each joint there is a transverse row of tubercles tipped with spines, the two stigmatal on each side dull ochre, the rest black.

The chrysalis is like that of *P. Cardui* and *P. Huntera*, but blackish varied with whitish, without metallic spots.

The larvæ feed on species of Gerardia, plantain, and snapdragon, and are to be found in the Middle and Southern States to the Pacific; occasionally in Massachusetts and Maine.

67. ANARTIA JATROPHÆ, Linn.

Expanse of wings 2 inches.

Color gray, with a livid tint, two round black spots to each wing, those on the hind wings in the position of the eye-spots of *J. Cænia*, but lacking the apical one on the fore wings, slightly pupilled. Outer border consists of three dull-brownish crenate lines, the outer on the edge of the wing. Wings crossed by several wavy lines the same color as the border, five of these being bars in the cell of the fore wings and at its end, with several shades along the costal region.

Under side paler, the spots and transverse marks as above, the outer margin washed with brown. Antennæ black, the club ferruginous.

This species is found only in Florida and Texas.

68. EUNICA MONIMA, Cram.

Expanse of wings 1.6 inches.

Upper surface dark brown, with a decided purple reflection. Across the apical portion of the fore wings are two oblique rows of whitish spots, the one across the end of the cell consisting of three spots, the outer or sub-apical of two. Fringes fuscous gray.

Under side brownish drab, the white spots repeated, and the space about these blackish brown. The hind wings are crossed by three brown, tortuous lines. Between the middle and the outer are six obscure brownish eye-spots, some black in the centre, some white, in two sets of three each.

Florida and Texas.

69. TIMETES PETREUS, Cram.

Expanse of wings 3 inches.

Upper surface bright fulvous red, costal edge of fore wings black, more prominent towards the apex. Both wings are crossed by three black lines, which are nearly parallel to the outer margin. On the hind wings the outer line is double, expanded on the costa, where the included space is white. The hind wings have a black border which sends a shade on the posterior angle of the fore wings and is shaded inward with brownish black about the anal angle. On the inner edge of the border is an ocellus at the end of the outer transverse black line, another elongate one at the anal angle, and a trace of a third farther towards the apex. Hind wings ending in a tail .6 of an inch long, and the anal angle prolonged into another .2 of an inch, the latter with some gray-blue and purple scales. Apex of fore wings produced, the angle below the apex prominent.

Under side brown, with a violet reflection, the lines darker brown, ashy at the anal angle. Body fulvous above, white beneath.

Indian River, Florida.

70. VICTORINA STENELES, Linn.

Expanse of wings 3.3 inches.

Male.—Upper surface dull black, marked by grass-green spots or markings. These consist of a row or band extending obliquely across the middle of the fore wings and the base of the hind wings, and a subterminal row common to both wings. The first row occupies about

FIG. 57.

Victorina Steneles, male (natural size).

the middle third of the fore wing and is broken up into oblong spots, but on the hind wing it is continuous, there being but little black between this and the base. Fig. 57 shows how this band is broken up into spots, consisting of two beyond the cell; then an interspace without a spot; the third in the upper median space, continued into the cell by a blunt conical spot, the vein separating them, and both rounded inwardly; the fourth occupies the lower two-thirds of the lower median interspace, somewhat boot-shaped, the foot extending into the cell; the band is continued to the hind margin by a single nearly quadrate patch, with the brownish submedian vein crossing it. This band continues across the hind wing to the inner margin, crossed by the brownish veins. The outer row consists of small spots between the veins, nearly round, the first four on the fore wings and the last two on the hind wings inconspicuous, the others varying in diameter from one-third to nearly the whole distance between the veins. Besides these bands or rows of spots there are two small patches along the costa, somewhat paler than the others. The black between the rows has a shade of drab across it, more prominent on the hind wings, it being orange-tinted near the anal angle.

Under side with the green repeated but much enlarged, so as to cover most of the under surface; the inner band bordered on each side on the hind wings and partially on the fore wings with orange, the outer band tinted on the inside with orange and brown. Ground color of the outer part yellowish drab, of the basal part almost white, the two shading into each other.

Fore wings considerably falcate, the apex slightly produced, the outer margin dentate; the hind wings more strongly dentate, with a tail near the middle .2 of an inch long; the excavation in margins of both wings white.

Female.—This resembles the male closely, but differs in the third spot of the median band not being accompanied by a spot in the cell, and in the under side having more orange.

Florida, Cuba, Mexico, Central America to Brazil.

18

71. DIADEMA MISIPPUS, Linn.

Expanse of wings from 2.5 to 3 inches.

Wings dark chocolate color, almost black, but when held against the light in a certain direction display a bluish-purple tint. The fore wings have a large white oval spot in the middle, and another smaller oblong white spot at the tips. The hind wings have a white spot larger than in the fore wings: on the edges of all these spots the purple color before mentioned is very conspicuous.

In vol. i. page 30, of "Papilio," Mr. W. H. Edwards quotes the description of this species from Westwood's edition of Drury, which is given substantially above, and states that a fine male of the species had been taken at Indian River, Florida, by Mr. Wittfeld, November 11, 1880, two others being seen at the same time but not captured. This is supposed to be the first time the species has been taken in the United States during the present generation, though by Drury it was credited to this country as far north as New York.

72. LIMENITIS URSULA, Fab.

Expanse of wings 3 inches.

Upper surface black tinged with bluish or greenish, and a little with fulvous at the apex of the fore wings. Along the outer margin are two rows of blue or green spots, the outer in the form of crescents, the inner lunules.

Under side brownish black, the outer border repeated, preceded by a row of black and a row of fulvous spots, some of the latter obsolete near the posterior angle.

There are two fulvous spots in the cell of the fore wings, three near the base of the hind wings, and some on the costa of both wings near the base.

The larva, according to Harris, is like that of *L. Disippus* in form, of a brownish color, more or less variegated with white on the sides, and with green above, and,

FIG. 58.

Limenitis Ursula (natural size).

like that of *Disippus*, has two long barbed brown horns on the second (third?) segment.

The chrysalis is not to be distinguished from that of *Disippus* in form and color, and the butterfly emerges from it in eleven days after the insect has entered this state.

The food-plants are willow, wild gooseberry, wild cherry, apple, plum, hawthorn, oak, *Vaccinium stamineum*, and *Carpinus Americana*.

Atlantic States, Mississippi Valley, Kansas.

73. Limenitis Arthemis, Drury.

Expanse of wings from 2.5 to 3 inches.

Upper surface black, a white band crossing both wings, that on the fore wings curving from the middle of the costa to near the posterior angle, from which place it is continued across the hind wings to above the anal angle. The fore wings have a white subapical spot and two marginal rows of faint green lunules. The hind wings have the lunules more distinct, and inside them a row of fulvous spots.

The under side is fulvous brown, marked as in *L. Ursula*, except that in this the white band of the upper surface is repeated.

This is a dimorphic species, the two forms at first supposed to be distinct species, but Mr. Edwards has reared both forms from eggs deposited by the same female, which proves them to be only forms of the same species. The above description applies to the one known as dimorphic form Lamina, Fab. The other, dimorphic form Proserpina, Edw., may be known by the white band of the fore wings being obsolete on its anterior half, or by there being at most only a whitish stripe occupying some part of the position of the band on the other form.

The egg is grayish green, dome-shaped, with the whole surface covered with six-sided reticulations, from each angle of which arises a short, tapering, white spine.

The young larva hatches from this in from seven to nine days. It is yellowish brown, covered with fine tubercles, each supporting a fine club-shaped hair. The young larva is said to eat away the leaf on both sides of the midrib, and when at rest is to be found on this stripped portion.

After the first moult the color is blackish brown, with a light brown patch on joint 8, covering the dorsum and part of the sides. On all the joints back of the second there is a broad ridge, in most cases followed by two narrow ones; the broad one on joint 3 elevated and bearing two tawny conical processes, crowned by a cluster of little fleshy knobs. Besides these there are other lower processes along the back. After the second moult the dorsal patch becomes pale buff, and extends partly over joints 7 and 9.

Five days after this moult each larva begins to make its hibernaculum, or case in which to hibernate, and three days later they close themselves in these cases, where they remain till the following spring. After they emerge from the hibernaculi they feed about two days and moult the third time, when they are red-brown speckled and mottled with black, with the processes ferruginous. Joints 2 to 4 are yellowish inclining to buff. After the fourth moult the color remains about the same.

The mature larva is 1.2 inches long; the red color two days after the fourth moult begins to change to green, olive, and partly a light and deep green; the dorsal patch to sordid buff, dull red buff, or whitish; the anterior segments to gray or whitish. The head changes from red to blue, and becomes dark drab.

The chrysalis is .9 of an inch long, subcylindrical, the abdomen somewhat compressed laterally and terminating rather abruptly; the general shape much lik· that of *L. Disippus.* The color of the wing-cases and anterior parts silvery gray, the former tinged with brown

or pale black along their hind margins; the wing-cases
varying somewhat in shade.   Dorsal side of abdomen
yellow-white, gray towards extremity; ventral side almost wholly gray and brown; dorsal appendage dark
smoky brown, with silvery corrugations before and after
it.   The butterfly emerges from the chrysalis from June
to July.

The food-plants are willow, aspen, basswood, and
probably thorn.

Northern United States.

74. LIMENITIS DISIPPUS, Godt.

Expanse of wings from 2.5 to 3 inches.

Upper surface fulvous, the lines black ; on the fore
wings a black triangular patch, containing three white

FIG. 59.

Limenitis Disippus,—right wings the under side (natural size).

spots on the costa beyond the cell, continued to the posterior angle in a narrow line.   A black line crosses the
hind wings about two-thirds of the distance from the
base, as shown in Fig. 59.   Outer border black, contain-

ing a row of white spots; and there are two anteapical white spots, the lower one quite small. Fringes black spotted with white.

Under side similar to the upper, but the fulvous is paler; the border contains two rows of white spots, and white spots between the veins in front of the transverse line of the hind wings, and on the costa of the fore wings in front of the cell.

Var. FLORIDENSIS, Strecker.—This form has the upper surface dark, almost mahogany color, but the

Fig. 60.

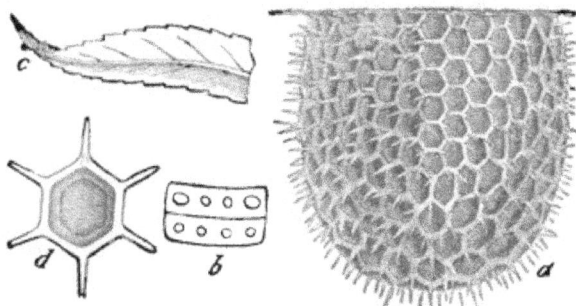

L. Disippus: *a*, egg, ×30; *c*, same, natural size; *d*, more enlarged view of one of the cells.

under side is as pale as the usual form. This form is found from the southern part of Illinois south.

Var. PSEUDODORIPPUS, Strecker.—On this the mesial black stripe of the hind wings is wanting; the anteapical black patch almost gone,—only indicated by a darkish shade devoid of the usual three white spots. Under surface same as above, save that in the submarginal row of white lunules there is no intervening black line between them and the reddish ground color. This was from a single female taken in the Catskill Mountains, New York.

The egg represented in Fig. 60 is similar in form and size to that of *L. Arthemis*. The eggs are pale yellow at first, but soon change to gray.

The young larva is yellowish brown mottled with dark streaks, especially below the stigmata; head twice as large as joint 2, and bilobed. Each joint is divided by a transverse impressed line, and on the top of the folds thus made are four elevated spots, the anterior the largest. There is also a subdorsal and substigmatal row of similar

FIG. 61.

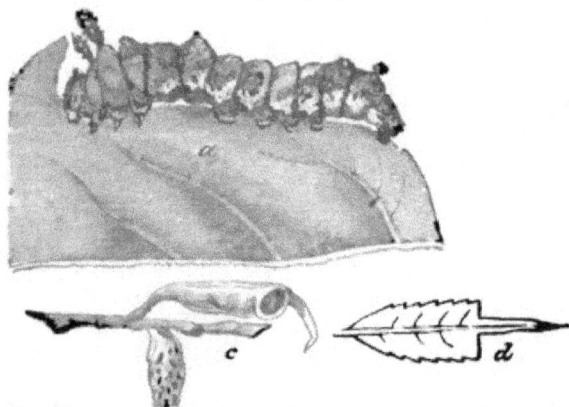

L. Disippus: *a*, mature larva; *c*, hibernaculum; *d*, leaf eaten all but midrib.

warts, from each of which springs a pale bristle. The second period scarcely differs from the first. In the third period the horns acquire their mature proportions, and the whole surface of the larva becomes more granulated. In the fourth or last period the blue points appear, and the lateral rows of tubercles lose their conspicuousness to a great extent.

The mature larva is 1.2 inches long; general color either whitish or olive-green. Body thickly granulated.

Head dull olive, with dense minute prickles; bilobed, upon the vertices a pair of prickly horns. Back specked and mottled above the stigmata with olive of different shades, except joints 3 and 9 and the upper parts of 8 and 10, but with a continuous pure white substigmatal line. Below this, on joints 5 to 11, is a large olive patch, on joints 7 to 10 extending to the tip of the prolegs. A pair of black, transversely-arranged dorsal dots in the sutures behind joint 3, and a more or less obvious lateral one just above, and behind the sixth and eighth pair of stigmata. Joints 4 to 8 and 10 to 12 with more or less shining, elevated, blue dots. On joint 3 is a pair of prickly, cylindrical, black horns, transversely arranged, .16 of an inch long; on joints 4, 11, and 12, a pair of dorsal tubercles, each crowned by a little bunch of from eight to twelve prickles; on joint 6, a pair of similar tubercles, but larger and of a yellowish color; on joints 5, 7, 8, and 10, tubercles similar to those on joints 4, 11, and 12, but smaller; on joint 13, four black, prickly, dorsal horns.

FIG. 62.

b

L. Disippus, pupa.

The pupa (Fig. 62) is similar in form to that of *L. Arthemis*, and is marked with burnt umber-brown, ash-gray, flesh color, and white.

The winter is passed in a hibernaculum consisting of a leaf, similar to *L. Arthemis*, except when there is more than one brood in a season. In this case it is only the last brood that has a torpid state.

The food-plants are apple, plum, willow, poplar, and oak.

United States generally.

75. LIMENITIS EROS, Edw.

Expanse of wings from 2.6 to 3 inches.

Upper surface dark red-brown, mahogany color, the black markings heavier than in *L. Disippus*, the white spots in the border of the hind wings subobsolete, while in the males there are often a series of white crescents in front of the black line of the hind wings.

The under side is as dark as the upper, or scarcely lighter, with the white marks more prominent than in *L. Disippus*, and a white spot near the base of each wing.

The following description of the preparatory stages, arranged from Mr. Edwards's account of the life-history of this species, will show how the two species differ in the larva state.

The egg is similar to that of *L. Disippus*, but a little higher in proportion.

The color of the larva after the first moult is mottled tawny and dark brown, the appendages on joint 3 one-third as large as in *Disippus* (.01 of an inch).

After the second moult, color more black, less tawny, the appendages .03 of an inch long, thick, club-shaped, covered closely with grains, mostly tawny, a few black.

After the third moult, color black, the tops of all the tubercles tawny, the appendages .06 to .08 of an inch long, clubbed as before, tawny.

After the fourth moult, color variable, either dark red-brown, the anterior segments brown-buff, the patch light buff, pink-tinted, or dark yellow-brown, the anterior segments yellow-white, the patch yellow, with buff tint; the appendages .12 to .22 of an inch long, clubbed and closely covered throughout with tawny grains (the shorter processes clubbed, the longer tapering, but clubbed at tip). Head amber color, in some cases yellow-brown, the top and sides pinkish.

In 1878, Mr. Herman Strecker briefly described, in his Catalogue of Butterflies, a Southern form as darker than *Disippus*, under the name Var. *Floridensis*, but gave no marks of difference except the color. In the December number of the " Canadian Entomologist" for 1880, Mr. W. H. Edwards, recognizing two forms in the Southern States as differing from *L. Disippus*, described the one farthest separated from *Disippus* as *L. Eros*, presuming that the other form was Mr. Strecker's var. *Floridensis:* the one described as *Eros* contained points of difference besides color, and Mr. Strecker's description made no mention of any other distinction. In a subsequent number of the " Canadian Entomologist" Mr. Strecker claimed that Mr. Edwards had redescribed his variety *Floridensis;* but the New York Entomological Club sustained Mr. Edwards in his name *Eros* for the form to which it was given. Following that decision, the two names are used in that way in Mr. Edwards's New Catalogue, and the nomenclature here is based on that arrangement.

The food-plant is willow.

Florida, Georgia.

76. APATURA CELTIS, Bd.—Lec.

Expanse of wings from 1.5 to 1.75 inches.

Upper surface russety gray or fawn color, inclined to olivaceous, shaded with black-brown. The outer half of the fore wings, except the hind margin, and two spots and a bar in the cell, dark brown, with an irregular row

of seven pale yellow spots beyond the cell. There are
three submarginal black ocelli, the lower not pupilled,
the middle pupilled, with the lower of three white spots ;
a little tawny near the posterior part of the margin.
Hind wings more uniformly shaded with brown, with
two marginal rows of fawn lunules; and beyond the
middle six round eye-like spots, not pupilled, in a pale
field. In certain lights the upper surface has a little
iridescence.

Under side light gray, less brown than above, the

Fig. 63.

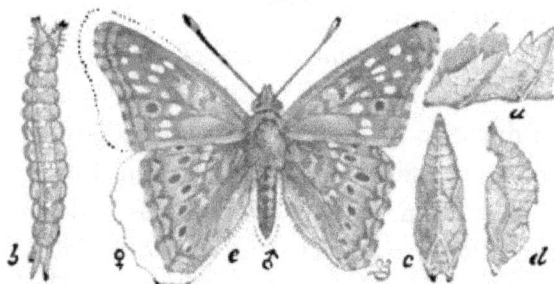

Apatura Celtis: *a*, egg; *b*, larva, dorsal view; *c, d*, pupa; *e*, imago, male, the
dotted line showing form of female.

middle of the fore wings with a slight yellow tinge.
The fore wings have two ocelli, the hind wings seven,
all annulate with pale yellow, and all but one on each
wing with a pupil, the pupil of those on the hind wings
pale blue.

The mature larvæ, as described by Professor Riley, are
rather more than an inch long, of a pea-green color, with
a series of yellow spots along the middle of the back,
and three yellow lines on each side, the intermediate one
undulating, often obsolete on the anterior part of each
joint, and containing a little lead-colored dimple. The

body is more or less thickly granulated with pale papillæ; swells in the middle, from which it tapers both ways, the anal extremity ending in two horns. The back and sides are flattened, the latter sloping slightly roof fashion. The most characteristic feature is the head, which, though variable in color, is always surmounted in this stage by

Fig. 64.

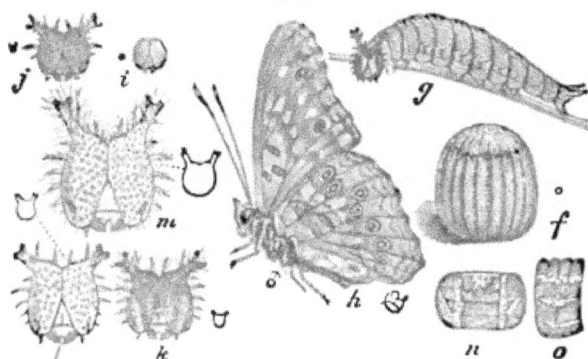

A. Celtis: *f*, egg, magnified; *g*, larva, lateral view; *h*, imago, under side; *i, j, k, l. m*, the five different larval heads; *n, o*, dorsal and lateral view of larval joint, enlarged.

two antlers. The heads in the different stages are well represented in Fig. 64.

The second brood of larvæ after passing the second or third moult cease to eat, station themselves on the under side of a leaf, shrink in size and change their color somewhat, and become torpid. In this state they hibernate.

Food-plant, hackberry.

Virginia to the Gulf of Mexico, Mississippi Valley.

77. Apatura Alicia, Edw.

Expanse of wings 2.25 inches.

Upper surface dull fulvous, a little more than the outer half of fore wings dark brown, except the hind margin

K   19

and a spur from the posterior angle. The cell has the three marks of the preceding species, also the row of spots beyond the cell, but the three next to the costa are white. Above the posterior angle is an eye-spot with a pale blue centre and a reddish annulus.

Hind wings with a border consisting of an edge and two black lines, with six ocelli within this border, about four of which are pupilled; and some shading of brown near the base.

Under side grayish white shaded with gray, the middle of the fore wings tinged a little with yellow. Marked much as above, the middle of the terminal lines yellow. There are two ocelli on the fore wings and seven on the hind wings, annulate with yellow and black, except the posterior one of the fore wings, which lacks the black, the anal one double.

Florida to Texas.

78. APATURA CLYTON, Bd.—Lec.

Expanse of wings from 1.75 to 2.5 inches.

Male.—Upper surface of fore wings rusty fulvous at the base, the remainder blackish brown, the veins sometimes ferruginous. The hind wings are blackish brown, the inner side with long greenish-brown hairs. Both wings have a black line forming the inner part of the terminal border, which is preceded by a series of rusty yellowish spots, obsolete at the apex of fore wings. Beyond the cell of fore wings are two rows of yellowish or rusty yellow spots, the outer row lacking two of reaching the hind margin; and there are two black bars in the cell. The hind wings have within the outer margin a series of six round black spots circled with ferruginous.

Under side of fore wings brown in several shades, gray in the cell; the marks of the upper side repeated, but not in the same colors, also a sinuous median black stripe. Hind wings purplish brown at base, limited through the middle by a darker sinuate line. Beyond this is a pale shade followed by another brown space containing the usual seven small ocelli. Outer border of both wings two crenate purplish-brown lines.

Female.—The fore wings are lighter, less brown on

FIG. 65.

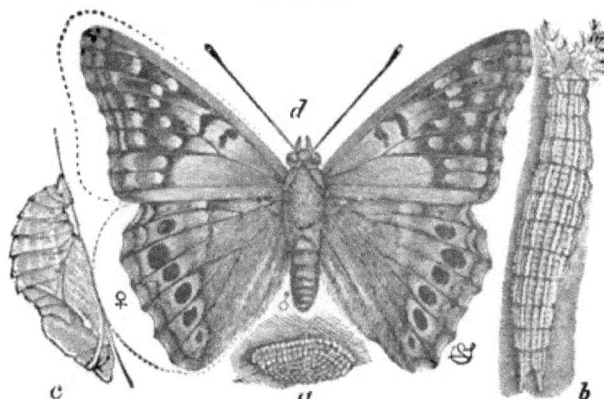

Apatura Clyton: *a*, eggs; *b*, larva; *c*, pupa; *d*, imago, male, the dotted line showing the form of female (all natural size).

the outer part, and the spots have a decided yellow tint. The under side in some examples quite pale.

This species is dimorphic, but not seasonally, as there is only one brood during a season. The description given applies to the dimorphic form OCELLATA, Edw. Dimorphic form PROSERPINA, Scud., differs from this in having the hind wings wholly blackish brown, except a little ferruginous at base, the round spots showing very little or not at all.

The under side is darker, with more of a purplish tint, and the ocelli of the hind wings are obscured.

The eggs of this butterfly are deposited in dense clusters, those of *A. Celtis* singly. When first deposited the eggs are pale yellowish white, but towards time for hatching the mass becomes more buff-colored.

The larva, in its first stage, is, according to Professor C. V. Riley, easily distinguished from that of *Celtis* by its copal-yellow, instead of black, head; and in the other

FIG. 66.

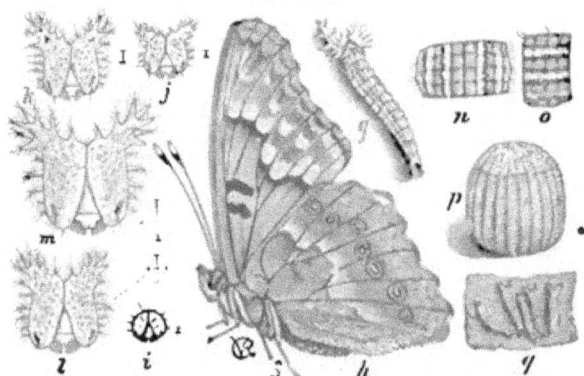

Apatura Clyton: *g*, larva, half grown, dorsal view; *h*, imago, male, under side; *i, j, k, l, m*, the five different heads of larva; *n, o*, dorsal and lateral views of larval joint; *p*, egg, enlarged; *q*, larvæ as when hibernating (natural size).

stages by a dark dorsal line, and a straight instead of wavy suprastigmatal line. The head is also larger, more pubescent, broader at the apex, and with the antlers larger, more spiny and hairy. These characters are well shown in the two illustrations which accompany each of the two species compared. According to Mr. W. H. Edwards, there is only one brood of *Clyton* in the latitude of West Virginia, the larvæ hibernating after the second or third moult.

The food-plant is hackberry, *Celtis occidentalis.*
New York to the Gulf of Mexico, Mississippi Valley,
Kansas.

79. APATURA FLORA, Edw.

Expanse of wings about the same as *A. Clyton.*

Male.—Both wings more excised than is usual in
Clyton, and the hind wings more prolonged and more
pointed at the anal angle. Upper surface of both wings
uniform bright orange ferruginous, except the area beyond the cell of the fore wings, which is of a deep shade
of ferruginous, blackened in the middle of the several
interspaces. The fore wings are scarcely at all obscured
at base, and the two rows of spots are bright orange
ferruginous, of the' same shade as the general surface,
instead of being yellowish as in the usual *Clyton.*

The hind wings have the base and inner margin but
slightly obscured, and a broad bright stripe extends from
the middle of the wing to the marginal band. The
ocelli lie in this field, and are large. The marginal band
of each wing is remarkably broad, so that on the hind
wings it nearly reaches the ocelli; and, except in the two
interspaces next the outer angle, there is a total absence
of the submarginal crenated line always seen in var.
Ocellata of *Clyton.* Furthermore, there is an absence
of the light patch on costal margin. The peculiar shape
of the wings, the uniform bright shade of ferruginous,
extending even to the rows of spots beyond the cell, the
large ocelli, the broad marginal band, and the absence of
the crenated line and of the costal patch, strike the eye
at once.

On the under side the pattern is as in var. *Ocellata*

of *Clyton*, but the colors are all intense; the cell and nearly all the spots of the fore wings buff, the extra discal area deep ferruginous; the basal area of hind wings deep gray-brown, tinted with ferruginous next the costa and towards the anal angle. The sinuous discal stripe is deep ferruginous, as is also the field on which are the ocelli, and between this stripe and field the space is lilacinous; the ocelli intense ferruginous, with obsolete rings, and lilacinous pupils. On both wings the broad marginal band is cut by a conspicuous blue-black stripe from the anal angle to the second subcostal venule on the fore wings. This stripe is so expanded next the posterior angle of the fore wings as to occupy full half the width of the band.

Female.—Duller colored, but as little obscured as the male. The fore wings are crossed by a broad, sinuous, deep black discal band, which in the usual *Clyton* is brown or ferruginous, and the bars in the cell are black and heavy. A broad submarginal black stripe completely crosses both wings, the margin outside this stripe being ferruginous, concolored with the cell. The crenated line is absent from the hind wings even at the outer angle.

This description is slightly modified from Mr. Edwards's description as given in "Butterflies of North America," vol. ii. The following description of the preparatory stages is from the same author in the "Canadian Entomologist" for May, 1881.

The egg is similar to that of *Clyton*, nearly spherical, flattened at base, marked by eighteen or twenty slightly prominent vertical ribs and many fine equidistant striæ: color yellow-green.

The young larva, which hatches in six or seven days, is .08 of an inch long, cylindrical, tapering from joint 3 ; pubescent, and of a pale translucent green color. The head is twice as broad as the second segment, subglobose, bilobed, the surface thickly pitted with yellow excavations ; color shining yellow or ochrey brown. Towards the last of the stage the body is less tapering, each segment well rounded, with dark green dorsal and subdorsal lines, the last more decidedly green. After six days the larva moults, when it is .14 of an inch long, the body a little thickened at joint 7, tapering slightly each way, the last segment ending in a forked tail. The surface is closely covered with yellow and yellow-white tubercles, arranged in longitudinal rows, and also in regular cross-rows. These tubercles are stout at the base, subconic at the top, of irregular size, and at the top of each is a short white appendage. Dorsal and subdorsal stripes dark green, the last narrow. The dorsum is covered by two bands of tubercles, divided by the green stripe, each band made up of two rows, the outer row whitish, the inner yellow. The subdorsal region, or below the lateral stripe, is another band of two rows, and as the stage proceeds these separate, showing a dull green line between them, the lower row running with the basal ridge of the body. The tails are divergent from base, short, tapering, rough with tubercles, and these give out longer hairs than elsewhere. Under side, feet, and legs yellow-green. Head subquadrate, the sides rounded, the whole surface shallowly pitted and covered with short yellow down ; color greenish white, with dark brown spots and patches, a large brown triangle over the mandibles, a small subtriangular patch at

the top in the depression, a subcrescent patch along the
base of each vertex, a stripe from the vertex half-way
down the side, and another down the back; the ocelli
black on a brown ground; on each vertex a short, compressed, fleshy, white process, and single white spurs
along the back of the head at the top and down the
sides; on the sides and top of the processes and spurs
many long white hairs.

After four days the larva moults again, when it is
.26 of an inch long; the shape similar but stouter, the
sides somewhat less rounded than the dorsum, the base
broadest, and the tails more produced. The tubercles
remain as during the preceding stage, but broader and
flattened, the two rows of each dorsal band a little separated, so as to show a dull green imperfect line; the side-
stripe much widened, joint 2 wholly yellow. The head
is shaped as before, the depression more angular, green
behind, green with dark brown patches at the sides and
front ; these patches much extended, the one at the top
nearly meeting by a triangular projection the triangle
from the mandibles, and the one from base of process in
nearly all cases protracted to outer end of mandibles, so
that the white in front is confined to two curved vertical
stripes, forming with a cross-line between the two triangles
the letter H. The processes are stout, short, cylindrical,
evenly forked at the top, each fork bluntly rounded, and
a little tapering; at the base is a spur turned forward,
and along the back and sides are single spurs; color of
process black in front, green behind. Along the back of
the head at the top are spurs in line, and so down the
sides, diminishing gradually in length, the upper ones
bent down ; all, as well as the processes, pilose.

It moults again in three days, but the characters do not change. Length from .38 to .4 of an inch.

In five days more it moults again, when it measures .6 of an inch, the shape and markings unchanged.

The mature larva is from 1.2 to 1.4 inches long, sub-cylindrical and robust, the dorsum well rounded, the sides much less convex, rather flattened, and sloping to a broad base; the last segment ending in a forked tail. It is banded with tubercles as at the first part of this stage, but these have constantly diminished in size as the larva progressed, and are in no place so distinct, and many have disappeared altogether. General color either bright yellow or ochre-yellow, a little whitened along the edges of the dorsal area; dorsal stripe very narrow, and either black or deep blue, but greenish on two or three anterior joints; the two side-stripes are dull or sordid green, as is also the under side. Head subquad-rate, the sides rounded, the front moderately rounded, the top depressed, on each vertex a short stout stag-horn process, with four prongs, these and the entire front and sides of the processes black; the back is green, and upon it and at the sides below are four green similar prongs, spurs similar to those in the former stage; the rest of the head greenish white and black, and thickly covered with a fine yellow down; the processes and spurs much covered with long white hairs.

The chrysalis is from .7 to .85 of an inch long, shaped much as in the other species. The color is pale yellow-green, finely streaked and speckled with light buff over head-case, mesonotum, and wing-cases, and speckled over abdomen. In six or seven days after pupation the butter-fly emerges. Like the other species, this feeds during the

p

larval state on hackberry, with habits similar to those of *Clyton*.

Texas, Florida.

80. PAPHIA TROGLODYTA, Fab.

Expanse of wings from 1.7 to 1.9 inches.

Male (Fig. 67).—Upper surface copper-red, outer margin edged with a black border, with gray or purple

FIG. 67.

Paphia Troglodyta, male (natural size).

reflections. At the end of the cell of the fore wings is a black bar, and a black patch on the costa of the hind wings.

Under side dead-leaf brown with a gray lustre, tinted on hind margin of fore wings with reddish, and throughout covered with dark scales. The basal half of fore wings deep-colored, limited without by a wavy edge; beyond this, and reaching to the dark marginal border, is a broad wavy band of paler color, bifurcate at the costal margin. There is a small cluster of luteous scales in the subcostal interspace of the hind wings and on the middle of the outer area.

Female.—This has the upper side pale red, the marginal border of fore wings very broad, enclosing a yellow-red wavy band imperfectly bifurcated. The hind wings have a similar band, contracted at the apex, and greatly expanded through the middle, which is partially sepa-

Fig. 68

Paphia Troglodyta, female (natural size).

rated from the apical portion by a line. The under side is vinous brown, with gray reflections.

The young larva is light bluish green thickly covered with soiled white papillæ. Scattered among these are light orange papillæ of a larger size, with occasionally one of brown. The head is larger than the third segment, which is the largest in the body. At each moult some of the papillæ disappear, especially the brown ones, the green shade becoming more apparent and the skin softer.

The mature larva is 1.55 inches long, cylindrical, tapering each way from the third joint. Color light bluish green; surface rough, covered with whitish papillæ. Head bilobed, a pair of orange papillæ on the

FIG. 69.

Paphia Troglodyta: *a*, larva; *b*, pupa.

vertex. Neck green, constricted, retracted within joint 2 when at rest.

The chrysalis is at first light green, soon changing to whitish green or to light cinereous brown; the whole surface indistinctly marked with fine parallel streaks of darker color. In form short, thick, gibbous, the abdominal joints contracted almost into a hemisphere.

The food-plant is *Croton capitatum*, and probably also *Croton monanthogynum*, as the butterfly is to be found where the first plant does not grow but the latter does. Western States from Illinois to Texas, Nebraska.

SUBFAMILY SATYRINÆ.

In this group the palpi are close, elevated, very hairy, the discal cells always closed, and the veins of the fore wings usually dilated at base. In Debis the eyes are hairy, in Satyrus naked, while in Neonympha they vary, some species being hairy and others naked. The butterflies in the Eastern United States vary from russet to dark wood-brown or nearly black, and in most species have eye-spots on the under side of the hind wings at least. The larvæ are cylindrical, tapering slightly from the second segment, the head larger than this segment. The body is more or less provided with small tubercles supporting hairs. Grass constitutes the principal food-plant. The chrysalides are more nearly cylindrical than in the preceding subfamily.

81. DEBIS PORTLANDIA, Fab.

Expanse of wings from 1.75 to 2.25 inches.

Upper surface wood-brown, rather light, the outer third a little paler, the division marked by a darker shade of the ground color, more pronounced on the fore wings, the line dentate with two prominent teeth opposite the discal cell, from which it bends inward before a whitish somewhat oblong costal patch. In this paler terminal space there is an anteapical whitish dot in line with four or five round dull black spots annulate with

20

yellowish, two of them small, the three larger pupilled
with dark black. On the hind wings there are five

FIG. 70.

subequal rather large
eye-spots. The outer
border is composed of
two lines slightly paler
than the ground color,
with a dark line and a
dark internal shade.

Under side brown as
above, with a violet
tinge, traversed by two
brown sinuous lines, be-

Debis Portlandia, male (natural size).

tween which there is a discoidal arc of the same color.
The eye-spots are brighter and blacker than above, the
iris yellow and pupil white, the anal one on the hind
wings double. Those on the fore wings are enclosed in
an oblong white ring. A similar ring enclosing those of
the hind wings is crenate, the first and the last cut off
from the others by cross-lines. The border is composed
of white, brown, and dark yellow lines.

This species is to be found in woodlands, the **male**
sitting on the body of some tree, from which it flies upon
the approach of any intruder. After flitting about the
trespasser upon its domain it returns to the same or an
adjacent tree. The females are mostly to be found on
the wild grasses that grow in such places, upon which
the larvæ feed.

The eggs are obovoid, the base a little flattened, and
under the middle thereof is a slightly rounded protuberance of less diameter, smooth. Color greenish white.

The young larva is cylindrical, head twice the diameter of any other joint, body tapering slightly from joint 2 back. Each segment from 3 to 12 is creased, making six ridges; on each of the first and fifth is a minute tubercle with a rather long hair bent forward, making two dorsal rows, with a similar row on the sides, and more lower down. Color of body whitish yellow, changing to pale green. Head slightly cordate, on each vertex a slight protuberance with a long curved hair, and other hairs over the surface. Color yellow.

After the first moult there is but little change, the body ending in two long, slender, blunt-tipped tails; color green, tubercles white. After the second moult the color is light green, the subdorsal tubercles more yellow, horns long, with red tips. After the third moult the color is the same. The larva hibernates in this stage, moulting twice more after reviving from its lethargy in the spring. After the fourth moult the color is yellow-green, with dark green dorsal and subdorsal stripes, and one below these, all narrow.

The mature larva is from 1.2 to 1.4 inches long, the dorsum much arched, and sloping about equally each way from the middle, ending in two small, short, slender tails. Each joint is creased, the first ridge broadest and bearing many fine whitish tubercles, mostly arranged in rows. Color yellow-green striped with dark green and yellow. Head yellow-green, the vertices bearing stout, tapering horns, red-tipped.

The chrysalis is .6 of an inch long, green, the ventral sides of abdomen whitish; top of head-cases and edge of wing-cases cream color, the surface smooth and glossy.

The butterfly emerges from the chrysalis in about fourteen days, appearing on the wing in July, specimens being seen from this time through most of the season. It occurs from Maine to the Rocky Mountains and to the Gulf of Mexico.

82. Neonympha Canthus, Bd.—Lec.

Expanse of wings from 1.75 to 2 inches.

Upper surface pale brownish gray, the outer third of both wings paler, the line of division between this paler portion and the basal distinct and dentate. On the fore wings this pale band contains three or four blurred ocelli; the hind wings having six, larger and more distinct. The outer border contains three lines, two light and one dark, besides the dark edge.

The under side is a little more brown-tinted, the darker shade defined on its outer edge by a brown line, another brown line cutting the inner third. The terminal band has the ocelli all annulate with yellow-buff, outside this buff a brown and then a whitish ring, the first and fourth on the fore wings and all on the hind wings pupilled with white. There are five ocelli on the fore wings and six on the hind wings, the anal one geminate. Border as above.

The eggs are subround, broadest at the base, where they are flattened, smooth; color greenish white. These are laid singly on the stems of grass, hatching in about seven days.

The young larva is at first yellow-white, changing later to pale green. It is cylindrical, long, slender, the last segment bluntly forked; on each joint a few tubercles, from each of which arises a clubbed white hair. The head is nearly twice as broad as the second segment, with a rounded prominence on each vertex, indented at the top with a little tubercle in the middle of the hollow, from which springs a bristle; color light brown.

After the first moult, which takes place in eight days, the more slender tails are pink-tipped, creased transversely, and on the creases fine white tubercles with short bristles. The color is at first greenish yellow, but afterwards changes to pale green.

Along the middle of the back is a dark green stripe free from tubercles, and on each edge of this is a line of white tubercles, another along the side, and a third along the base; between the last two are two other white lines. The head is yellowish green, with the surface finely tuberculated. On each vertex is a long, tapering, rough horn, tipped with brown, and marked in front by a reddish stripe, which is extended down the side of the face.

In nine days it moults a second time, when it is .4 of an inch long, of the same form as before, and yellowish green, with the same tuberculated lines. In fifteen days it moults a third time, when it is .55 of an inch long, of the same form and color as before, but very soon changes to brown and buff. Along the middle of the back is a brown stripe, on each side of which is a reddish buff band, which changes to greenish on the outer side. There is another buff band on the side, through the middle of which runs a brown line. The basal ridge is buff. In a few days the larvæ become lethargic, and in this state they pass the winter. Soon after their revival in the spring they moult the fourth time, when they are .62 of an inch long, pale green, with a dark stripe along the back, and a yellowish white one just below; the two lines on the side and the basal stripe of the same color. Tails green; head emerald-green, the horns reddish, the stripe down the face dark brown. In thirty days they pass the fifth and last moult.

In some cases they hibernate after the fourth moult, when the color is green; but twenty-four hours after it changes to yellow-buff and red-brown, the dorsal stripe pale brown; the bands on each side of a greenish yellow, the side brown, with a dull green line running through it. The head has the face green, the stripes reddish brown.

The mature larva is 1.2 inches long, slender, the dorsum arched; the last segment ending in two long, slender, conical tails, which are rough with tubercles. The joints are creased with six ridges to a joint, the whole surface finely tuberculated, a fine hair arising from each. Color of body green, dorsal stripe darker, on each side of this a pale green band, on the outer edge of which is a yellow-green stripe. On the side a pale green band divided by a yellow line. Head with a long, conical, tapering process from each vertex, the whole head rough with fine tubercles. Color of head yellow-green, the horns red, a brown stripe down the sides.

In a recent communication Mr. W. H. Edwards gives the following description of the chrysalis: Length .62 of an inch, slender; much resembles in shape the chrysalis of *Debis Portlandia,* the head-case more produced than in that species, bevelled to an edge transversely, excavated at the sides; mesonotum carinated, sides flat, apex almost angular, a very little rounded; color green; top of head-case and dorsal edges of wing buff; a buff mid-dorsal stripe, with a buff subdorsal stripe on each side; also a faint lateral stripe of the same color.

Northern States.

83. Neonympha Gemma, Hüb.

Expanse of wings 1.25 inches.

Upper surface gray, with a little wood-brown tint, no marks except two darker shades near the middle of the outer edge of the hind wing.

Under side about the same color, sprinkled with buff scales, the outer edge with a slight golden reflection. On the fore wings are three not very distinct, wavy, brown, transverse lines. On the middle of the outer surface of the hind wings is a large oval patch composed of white and reddish-brown scales, giving the patch a slight violet tint when seen without a glass. In the outer edge of this patch, standing on the intervenular spaces, are four roundish, vandyke-brown spots, on each an anchor-shaped spot of pale, metallic, bluish scales. The margin of the wing towards the apex and anal angle has a border of metallic, bluish scales, with dentations up the sides of the veins. There are two brown lines on the hind wings, besides a brown bordering to the terminal patch.

The eggs are globular, seemingly smooth, but under a high magnifying power are seen to be reticulated in irregular hexagons, the ridges flat and broad, having at the bottom of each depression a white point. Color yellowish green. These are deposited on grass, and hatch in from three to six days.

The young larvæ are .12 of an inch long, cylindrical, a little thickest in the middle, ending in two divergent tails, the point blunt and tipped with a white bristle. Color white, with white scattering hairs, after a few days changing to alternate stripes of white and green. Head subpyriform, one-half broader than the second segment,

broader than high, with a slight angular depression at the
summit; at each vertex a straight, round, divergent horn,
thick at base, pointed at top. Color of head and horns
blackish brown. It moults the first time in from six to
nine days, when it is a little thicker in the middle, the
tails more slender and brown-tipped. Each joint is
several times creased, and on the summit of each ridge
is a row of white tubercles with white hairs. Color
dark green, banded and striped longitudinally with white.
Head higher, brown, green-tinted in front.

Moults the second time in from five to ten days, when
it is .34 of an inch long; shaped and striped as before.

Moulting again in from five to eight days, the length
is .55 of an inch. The color is reddish buff, the summer
and fall larvæ with a dorsal band of dark brown; the
subdorsal stripe reduced to a broken dark brown line,
distinct only at the extreme hinder end of each segment
from 3 to 10; in the middle of each side a dark
brown stripe, at the base a rounded ridge, buff-colored,
and below this a brown line. Head green in front, with
brown stripes, back part and horns gray-brown. The
spring larvæ after the third moult are bright yellow-
green striped with dark green, tails pink-tipped. They
remain in this stage from five to six days.

The chrysalis is .52 of an inch long, cylindrical,
thickest through joints 6 and 7, shaped much as in the
preceding species. The color of the dorsum and entire
abdomen of the summer and fall broods is sordid yellow-
buff, the wing-, antennæ-, and palpi-cases more yellow,
the surface finely streaked with brown. In the spring
brood the color is blue-green, the dorsum streaked irregularly with whitish green.

There are three broods of the larvæ in a season, each moulting only three times in coming to maturity.

West Virginia to Gulf of Mexico, Southern Illinois.

84. Neonympha Areolatus, Sm.—Abb.

Expanse of wings from 1.3 to 1.5 inches.

Upper surface wood-brown, somewhat grayish, the border-lines faint, and without eye-spots.

Under side scarcely paler than above, but sprinkled with buff scales. Fore wings with three small dark brown spots narrowly circled with yellow and containing a few silvery scales. Hind wings with five various-shaped elongate spots, the long diameter with the length of the wing, each with bluish metallic scales, sometimes in a bunch, sometimes a buff centre with metallic points around it; each spot circled with buff. There is a common, dull, dark yellow line a little in front of the middle of the wings, another just touches the end of the cell, and a third is submarginal, the last two nearly meeting at their ends on each wing, forming a broad band without change of ground color, in which the ocelli are placed; the margin of the wing of the same color. On the fore wings these lines are not so distinct as on the hind wings.

The eggs are nearly globular, smooth, but under a high magnifying power are seen to be thickly covered with shallow depressions. Color pale green.

The young larva, which hatches in six days, is cylindrical, the last joint bluntly forked. Color delicate green. Over the body are many white hairs, and among these are black clubbed hairs in longitudinal rows. Head about twice as wide as any joint, a little depressed at the top, upon each vertex a short semiovoid appendage tipped with two divergent black hairs; color black. Moults after eight days.

After the first moult the body is a little arched; tails longer, faintly red. Color of body green, surface thickly covered with fine yellowish tubercular points; the back of the head and triangle over the mandibles green, the rest red-brown. It is now .22 of an inch long. It moults again in nine days, with a length of .3 of an inch, but the color does not change. It moults three times in coming to maturity.

The mature larva is from 1.1 to 1.3 inches long, joints 3 and 4 creased and divided into five ridges, the rest into six; surface thickly covered with small sharp tubercles; tails reddish, slender. Head reddish, on each vertex a little conical process.

The chrysalis is from .48 to .54 of an inch long. The color is green, the edges of the carina, wing-cases, and top of head-case cream color; some points and patches of whitish. The butterfly emerges from this in about ten days.

The food-plants are grasses, mostly of the coarser kinds.

Gulf States; occasional in New Jersey.

85. NEONYMPHA EURYTRIS, Fab.

Expanse of wings from 1.5 to 1.75 inches.

Upper surface brownish black or wood-brown; near the outer margin of each wing two eye-spots circled with yellow, the small pupils often double. The anterior ocellus of the hind wings is often obsolete, and the posterior is often supplemented by a small one at the anal angle. There are three dark lines forming the outer border,—one on the edge and two a little within the edge ; and in some specimens the ocelli are in a band a little paler than the ground color.

The under side is drab-gray, both wings crossed by two rusty lines ; the ocelli more prominent than above, each of the ocelli of the hind wings repeated, with the smaller one at the end of the series, and silver traces of two more between ; the triple terminal lines as above.

The egg is of a yellowish-green color, nearly round, covered with fine irregularly hexagonal reticulations.

The young larva hatches from this in eight days. It is .08 of an inch long, cylindrical, tapering each way from the middle, the last joint ending in two short tails, with fine white hairs on the body. Color pinkish white, marked longitudinally by seven crimson lines. Head subglobose, nearly twice as broad as any other segment, depressed slightly at the top, with a small conical process from each vertex ; color dark brown.

In seven days it passes the first moult, when it is .16 of an inch long, and changes to drab, of either a green or a red tint, with five dull red stripes, the whole surface finely but roughly tuberculated, each tubercle emitting a short hair. Color of head yellowish finely mottled with red. In six days it moults the second time, when it is .24 of an inch long, shaped as before, with a fleshy ridge at the feet. Color dull ochrey yellow, striped with seven brown stripes ; head yellow marked with brown. It feeds fourteen days before moulting the third time, when it is .44 of an inch long. The color is the same marked with reddish, the stripes subobsolete. It is thirty days before it moults the fourth and last time.

The mature larva is 1 inch long, flat at base, the dorsum rounded, the last joint forked. Color of dorsum yellow-brown, the sides darker; a dorsal brown band, and on each side of this, from joint 4 to joint 11, is a series of dark brown patches. The subdorsal area is separated from the dorsal by two wavy parallel lines, the upper dark, the lower yellowish, and on the side of each joint from 5 to 11 there is an oblique dark stripe. Basal ridge yellowish, the tails tipped with red. Head yellow-brown, with a small conical process on each vertex. The whole surface of the body is covered with sharp tubercles of irregular size, each emitting a short brown hair.

The chrysalis is .5 of an inch long, shaped as *Sosybius*; cylindrical; the abdomen stout, and larger than the anterior portion. Color pale yellow-brown, the wing-cases and anterior parts streaked with fine, abbreviated brown lines; brown spots on the wing-cases, the abdomen with two brown stripes and two rows of brown dots. The butterfly emerges in eleven days.

There are probably two broods in a season, the last brood of larvæ passing the winter in a torpid state. They feed upon grass, the eggs being deposited singly upon the blades. They are very sluggish at all times, and frequently pass days without eating.

Atlantic States, Mississippi Valley, Nebraska.

86. NEONYMPHA SOSYBIUS, Fab.

Expanse of wings 1.35 inches.

Upper surface wood-brown, somewhat grayish, the border-lines faint, and both wings without ocelli.

Under side a little paler than the upper, two brown lines

crossing the wings, between which is a brown dash at the
end of the cell of each wing; border-lines three, the
inner crenate. Near the outer end of the fore wings,
below the costa, is a black eye-spot annulate with yellow
and pupilled with pale blue; there is also a small one between this and the costa, and there are traces of three more
towards the hind margin. The hind wings have two
distinct ocelli, with a small one before the first, another
after the second, and traces of two more, with silver,
between the two.

The egg is semiovoid, the base flattened, and the
sides at base rounded; the surface under a low power
appears smooth, but under a high magnifying power it
is seen to be covered with shallow thimble-like depressions. Color greenish white.

In four days a larva .09 of an inch long hatches
from this. The shape is cylindrical, marked by five
or six longitudinal tuberculated ridges, each tubercle
emitting a clubbed white hair. Color white. Head
much larger than the second joint, bilobed, the vertices
without processes, black.

After six days it moults the first time and is .2 of
an inch long, cylindrical, tapering both ways from the
middle, the last joint forked. The color is light green,
the tubercles white, with three green stripes; legs and
under side green. After seven days it moults the second
time, the color being blue-green, the head yellow-green.
In six days it moults the third time, being then .42 of
an inch long, stout; color pale green, with the tubercles
white, head emerald-green. It moults four times in
coming to maturity.

The mature larva is .76 of an inch long, cylindrical,

thickest in the middle, the last joint ending in divergent tails. The color is emerald-green, much covered with fine yellow tubercles placed on ridges, caused by the creasing of the segments, and with larger tubercles placed in longitudinal rows, each emitting a white hair. Head round, broader than high, bilobed, covered with yellow, conical, fine points.

The chrysalis is like that of *N. Eurytris.* The color is green, on the abdomen yellow-green; on each side of dorsum of abdomen is a small ridge, and on each side of this are three black dots placed in pairs. On each side below the wing-cases is a brown stripe; keel of mesonotum brown, with brown mottlings on the wing-cases. The imago emerges in thirteen days.

The eggs are deposited on grass, and there are two broods in a season. The butterfly is found with *Gemma* and *Eurytris* within the edge of the forest, or, if in the open country, always near timber.

Middle and Southern States, Mississippi Valley.

87. Satyrus Pegala, Fab.

Expanse of wings 2.5 inches.

Upper surface blackish brown, a broad buff band on the outer part of the fore wings, not reaching either margin, and containing a single pupilled ocellus on its anterior end. In some female specimens another ocellus is found on the posterior end of the band, but more often the second ocellus is represented in both sexes by a black dot or a small round spot. Near the anal angle of the hind wings is a single black ocellus.

Under side brownish gray, both wings crossed by many abbreviated brown lines. The buff band and ocelli of the fore wings repeated, the hind wings with six ocelli in two sets of three each.

The larva is said to be gray, with one broad and one narrow white band. The food-plant is coarse wild grass.

Gulf States; occasional in New Jersey on the coast; Mount Holly, N.J.

88. Satyrus Alope, Fab.

Expanse of wings from 1.75 to 2.5 inches.

There are two dimorphic forms and five varieties and sub-varieties of this species, being mostly local instead of seasonal. It occurs from Canada to the Gulf of Mexico, and from the Atlantic to the Pacific: in different parts of this region the different forms breed true to their type, but on the border-land between these different localities there are numerous intergrades connecting them all. The principal form found in the Atlantic States from North Carolina to New York is *Alope;* the form *Nephele* in its typical markings occurs in Canada, with intergrades in New England and other bordering territory; *Maritima* occurs on Martha's Vineyard and Nantucket Islands; *Texana* is found in Texas; the one found from Illinois to the Rocky Mountains is *Olympus;* and the Pacific slope is represented by *Boöpis* and *Incana.* The following descriptions of such forms as occur in the Eastern United States, as well as of the preparatory stages, are abbreviated from Edwards's "Butterflies of North America."

Dimorphic form, Alope, Fab.—Male.—Upper surface blackish brown, darkest over the basal area; outer margin consisting of two fine parallel lines, a little within which is a black stripe. The fore wings have a transverse yellow band beyond the cell, sometimes a little ochraceous, and often more or less encroached upon by the brown ground. On this area are two ocelli, round, black, of variable size, and with or without a central point, which is white with blue scales. Behind the cell is a blackish indistinct sexual dash in the males. The hind wings have a small ocellus in a yellow ring near the anal angle (often wanting).

Under side yellow-brown; the band enlarged and of a paler color; the ocelli repeated, enlarged; the marginal lines distinct; the brown area covered with abbreviated darker streaks, which over the base and disks form somewhat concentric broken rings, limited without by a common dark stripe. On the fore wings this stripe borders on the yellow band; on the hind wings it is irregularly sinuous from margin to margin, throwing out a rounded prominence against the cell, followed by a rounded sinus on the median interspace. Across the middle of the cell, and below it, a dark stripe; the extra discal area less streaked. The ocelli vary from none to six, the full number being most often present, disposed in two groups of three, the middle one of each group the largest; all black, rounded, in narrow yellow rings, and with white dots in the centre edged by blue scales.

Female.—This differs from the male in the band being generally broader, clearer, and well defined on both edges, the ocelli well developed, with occasionally additional black points on the hind wings corresponding to the ocelli on the under side. A larger percentage than of the males have no ocelli on the under side of the hind wings.

New York to North Carolina.

Var. MARITIMA, Edw.—This form is of small size, dark color, and the band on the fore wings reddish yellow.

Martha's Vineyard, Long Island, Eastern New Jersey.

Dimorphic form, NEPHELE, Kirby.—Male.—Upper surface blackish brown, the marginal lines and stripe as in *Alope*, but often obsolete. The fore wings have two small black ocelli, placed as in *Alope*, without ring or band, but in some examples a faint yellowish shade represents the ring; sometimes a white dot in each ocellus, but usually the lower one is blind. Hind wings with or without ocellus, but if present it is blind and without ring.

Under side dark yellow-brown, faintly streaked as in *Alope*, but with less distinctness, and in many examples the discal stripe is nearly or quite obsolete, especially on the hind wings. The ocelli of the fore wings are enlarged, the rings distinct but not clear yellow, being dusky, or more or less obscured. The hind wings have small ocelli in fine russet rings, with central white dots and a few bluish scales; the number varies from none to six, but the largest proportion have six, and very few have less than three.

Female.—This has the upper side lighter and more brown; the area which in *Alope* is occupied by the band is of a slightly paler shade than the rest of the wing. The ocelli are large, with white centres and blue scales, and either without rings or imperfectly ringed with dusky yellow, the hind wings either with or without an ocellus. The under side is paler, the area of the band a little lighter than elsewhere, the rings sometimes quite

bright, but oftener dull or obscure yellow. The ocelli
of the hind wings rarely six, more often less than three,
sometimes none.

New England, New York.

Var. OLYMPUS, Edw.—This equals in size the typical
form. The males are almost black. The ocelli are very
small and without rings, but in some examples there is
a faint russet or yellowish tint about them, and perhaps
on the space between them. On the under side the rings
are russet or ochraceous on both wings, and there is
a perceptible bronzy reflection on the under side of the
hind wings, especially in the males.

Illinois to the Rocky Mountains.

The egg is conoidal, rounded at base and somewhat
flattened, truncated, the sides slightly convex; marked
by about eighteen vertical ridges, crossed by numerous
fine low ridges; summit rounded, covered with shallow
cells. Color lemon-yellow. It hatches in from fourteen
to twenty-eight days.

The young larva is .1 of an inch long, the head
considerably larger than the other segments. The body
is carnation, with seven crimson longitudinal lines, the
head light lemon-yellow specked with brown; sparsely
hairy. On the body there are six rows of long white
bristles, a part of which point forward and a part backward.

The larvæ become lethargic soon after leaving the eggs,
and thus hibernate. As soon as they begin to feed in the
spring the color changes to pale green with the lines red,
but later the ground color becomes deep green and the

stripes darker green. After the first moult the larva is
.16 of an inch long, cylindrical, the anterior part the
thickest, the body ending in two conical tails covered
with tubercles and bristles. Each joint is creased, the
ridges bearing many tubercles with long white hairs.
Color pale green, with seven green stripes, basal ridge
pale yellow, tails reddish. After twenty-three days it
moults the second time, when it is much as before.
Length .3 of an inch, the color yellowish green, with the
basal side yellow, and the stripes dark green. In fourteen days it passes the third moult, the length being
then .44 of an inch. It moults four times before reaching maturity.

The mature larva is 1.25 inches long, cylindrical,
robust, thickest in the middle, with two sharp, conical, .
divergent tails. Each joint is crossed by five or six
creases, the ridges covered with fine white papillæ, each
supporting a long or short white hair, if long appressed
to the surface. Color yellow-green, varying, on some
the dorsum more yellow; a dorsal green stripe and a
basal yellow one, with sometimes a faint yellow lateral
line. Head vivid green. It takes fourteen days to pass
from the third to the fourth moult, and
twenty-eight days more to the time when
the larva ceases feeding and pupates.

Fig. 71.

The chrysalis is a little more than half an
inch long, cylindrical, the abdomen conical,
the wing-cases slightly raised at the margins; head-case short, roundly excavated at
the sides, the top narrow. The male is one

S. Alope, pupa.

shade of either yellow-green or deep green, covered with
smooth specks and patches of lighter color, with several

cream-colored lines. The female has three longitudinal yellowish bands. The butterfly emerges in fourteen days.

The larva feeds on meadow-grass, and the butterfly is found mostly in the open fields, differing in both particulars from *Pegala*.

89. Chionobus Jutta, Hüb.

Expanse of wings from 2 to 2.25 inches.

Male.—Upper surface wood-brown or grayish brown ; the fore wings with an oblique shade below the cell, and with a subterminal row of six yellowish spots, the first, third, and sixth small, with or without black central points, the others with each a round black spot. Hind wings with four yellowish patches more or less shading into the ground color, the anal one with a small black spot, and traces of one in the patch next to this.

Under side of the fore wings much as in *C. Semidea*. Color about as above, crossed by a great number of abbreviated dark brown lines, obscure on the fore wings, except along the costa and near the apex, where the brown is more distinct and alternate with gray. About five of the yellowish patches are visible, only two of them distinct, and these have round black spots pupilled with white. The hind wings are much darker than the fore wings, and the dark marks are not distinctly defined ; sprinkled with white scales over the basal third, and an irregular band of these beyond the cell and along the outer margin ; a submarginal row of intervenular white points.

Female.—Upper surface of fore wings as in the male, only the yellowish patches are expanded and somewhat blended. On the hind wings, instead of the subterminal row there is a broad terminal suffusion of yellowish and brown, with mere traces of an anal spot.

Under surface as in the males, except that the hind wings are lighter, the dark brown lines more clearly defined and alternating with grayish and yellowish brown, the bands and points of white as in the male.

This species occurs in the northern parts of both Europe and America, but Orono, Maine, is the only place where it has been found in the United States. The larva is unknown, but it is probable that it feeds on Carex.

90. CHIONOBUS SEMIDEA, Say.

Expanse of wings 1.9 inches.

Upper surface clear wood-brown, the darker markings of the under side showing through a little; costa spotted with white, dark brown, and pale gray-brown. Fringes pale whitish towards the apex, widely cut with brown at the ends of the veins.

FIG. 72.

Chionobus Semidea, under side.

Under side about the same color as above, the fore wings traversed by a great many abbreviated lines, which are obscure dark brown, except on the costa and apical portion, where the marks are dark umber-brown alternating with white. The marks are somewhat gathered into bands just beyond the cell. Hind wings much as the fore wings, only that the dark lines are all clear dark brown, inclined to be gathered into bands through the cell and beyond it; the alternating light spots before the first and beyond the last are white or whitish, making two irregular whitish bands, with more sprinkling of white along the veins beyond the outer whitish band. Antennæ annulate with brown and white, the knob fulvous.

The larva is said to feed on *Carex rigida.*
White Mountains, N.H.

SUBFAMILY LIBYTHEINÆ.

The insects of the one genus belonging to this subfamily are readily known by their long, beak-like, porrect palpi, and by the males having four feet adapted for walking, while the females have six.

91. LIBYTHEA BACHMANNI, Kirtl.

Expanse of wings 1.8 inches.

This species is readily known by the beak-like palpi, which are about three times as long as the head, and the quadrate apex to the fore wings. The upper surface is brownish black on the fore wings; the lower half of the cell, and the space below that, about twice as large, are fulvous, with a patch of the same at the end of the cell of the hind wings. The fore wings have a white anteapical patch, at the end of the cell an oblique white bar, with a quadrate white spot beyond its posterior end.

The under side is brown, the fulvous area enlarged, the hind wings and tips of the fore wings lilac-tinted, the white spots repeated. The under side of the male is clouded with cretaceous.

The egg is of a pale green color, in shape an oblate spheroid, somewhat pointed at the base, and a little truncated and depressed at the summit, marked by eighteen or twenty longitudinal ribs, crossed by corrugations. This hatches in about four days.

The young larva is .04 of an inch in length, cylindrical, each joint with four transverse creases; covered with fine short down. Color green, semi-translucent. Head twice as broad as joint 2; obovoid; color yellow-brown.

After the first moult, which takes place in two days, the length is .08 or .09 of an inch. In two days more it passes the second moult, and the length is .12 of an inch. Upper side dark green, a faint shade of yellow over and along the basal ridge, rather macular. In two days more it passes the third moult, when the length is from .26 to .28 of an inch. The color is dull green, yellowish along and over the basal ridge, specked with white or yellow-white as before this moult. In two days more it passes to the fourth moult.

The mature larva is from .7 to .9 of an inch long, cylindrical, thickened at joints 3 and 4, the dorsum of the last segment abruptly curved down to the end. Color dark green, the lower side, and also the feet and legs, pale green. It is creased as in the preceding stages, and on the ridges are pale or whitish yellow points. Along the sides is a supra-stigmatal stripe, above which the color is yellowish for a little way, also a narrow dorsal line, sometimes a subdorsal one. Head ovoid, smooth, green.

The chrysalis is half an inch long, compressed; head-case square, mesonotal process rounded. Color blue-green, a faint yellow dorsal line from the last joint to the middle, where it forks, sending a branch over each wing-cover to the tip of the head-case, where they meet; a lateral abdominal yellow line. The abdomen marked with many white tubercles.

This butterfly differs from most others in that the males have only four feet developed for walking, while the females have all six fully developed. The eggs are deposited in the developing ends of twigs of *Celtis occidentalis* on the partially-developed leaves, only one to a branch. The time for reaching maturity after the fourth moult is four days, and the pupal period is from five to seven days.

Atlantic States, Mississippi Valley, Texas, Arizona.

FAMILY ERYCINIDÆ.

This family is represented by small or medium-sized butterflies, with the same arrangement of feet as the last subfamily of the Nymphalidæ, or the males with four feet and the females with six; but they may readily be known from the Libytheinæ by the palpi, which are short or only of moderate length. Little is known of the preparatory stages, but, according to Mr. H. W. Bates, "the metamorphoses are variable, some genera resembling the Nymphalidæ, in the chrysalis being suspended by the tail, and others the Lycænidæ, in being recumbent and girt with a silken thread. Too little is known of the caterpillars to enable us to say whether they offer any peculiarity." The preparatory stages offer good bases of classification, often showing the relation one group bears to another even when the imagines do not show this so clearly. The feet indicate that this family should follow Nymphalidæ, while the preparatory stages of *Lemonias Nais* show that they should come next to the Lycænidæ on the other side. Nearly all the species are found in America, mostly in the tropics.

SUBFAMILY ERYCININÆ.

92. CALEPHELIS CÆNIUS, Linn.

Expanse of wings .8 of an inch.

Upper surface ferruginous, the wings crossed by four or five more or less sinuous blackish lines, almost separable into spots. Beyond these are two lines of shining black-lead color, the inner sinuous; a row of black points between them.

Under side yellow fulvous, the lines and spots more distinct. Fringes blackish.

Southern States.

93. CALEPHELIS BOREALIS, Gr.—Rob.

Expanse of wings from 1 to 1.2 inches.

Upper surface umber-brown, shaded a little with reddish, both wings with four transverse sinuous lines of dark brown, the space along the outer row darker-shaded, near the outer margin two metallic lead-colored lines, with a row of dots between, the inner line sinuous. Outer margin reddish.

Under side yellowish orange, with the rows and spots of the upper surface repeated, but with less distinctness, except the metallic lead lines.

This species has been found in New York, New Jersey, West Virginia, Michigan, and Illinois.

22

94. Eumenia Atala, Poey.

Expanse of wings from 1.6 to 1.8 inches.

Male.—Upper surface black, the fore wings washed with green three-fourths the distance from the base to the apex. Hind wings with a little green on each side of the median vein, and with a broad border of inter-venular blue-green lunules.

Under side black; the fore wings without marks; the hind wings with three rows of small blue-green spots, the outer two parallel with the outer margin, the inner sinuous, just beyond the cell. On the inner margin is a large, somewhat oblong, reddish-orange patch, dentate towards the base, extending from the middle of the margin to near the base. Fringes black.

Body black; abdomen, except a portion of the base above, orange.

Female.—This differs from the male in having the green scales sprinkled throughout the discal cell of the fore wings and sometimes below this; the border lunules of the hind wings slightly yellowish green; the spots on the under side yellowish green; and the apex above and below bluish; fringe white.

Florida.

FAMILY LYCÆNIDÆ.

These are mostly small butterflies, having six feet adapted for walking. The larvæ are more or less oblong-oval or oval, with the head retractile into the second segment, and a dorsal tuberculous ridge. The chrysalis is short, attached by the anal extremity, with the tip of the abdomen bent forward and the body girt about with a silken thread much as in the Papilionidæ, but closer to the object to which it is attached. The number of feet and the manner of attaching the chrysalides would imply close relationship with the Papilionidæ, and some authors place them next to that family; but the head of the adult closely appressed against the body, the larvæ, by their shape, retractile heads, short feet, and manner of moving over a surface, suggesting the Limacodes group of the Bombycidæ, and the shape of the eggs somewhat like those of the Hesperidæ, all seem to indicate that they should be placed below the Nymphalidæ and the Erycinidæ. There are two subfamilies, Theclinæ and Lycæninæ.

SUBFAMILY THECLINÆ.

These generally have a rather stout body, the palpi very short, and the antennæ reaching to the middle of the fore wings; often the anal angle more or less produced, with one or two slender tails projecting from the outer margin near the anal angle.

95. THECLA HALESUS, Cram.

Expanse of wings from 1.4 to 1.6 inches.

Male.—Upper surface bright glossy blue, the outer fourth black, except towards the anal angle of the hind wings, where the border becomes narrow; a border of about the same width on the inner margin; a black sex-mark in the cell of the fore wings. Hind wings with two black tails, before the larger of which is a blue lunule, with a smaller one each side; the anal angle produced.

Under side brown-black, with a red spot at the base of

the fore wings, and two more at the base of the hind wings, the fore wings with a dash of blue along the median vein. The hind wings have the blue lunules repeated, with three others standing before these, yellow-green in a black ground.

Body blue-black above, black beneath, with the under side of abdomen orange.

Female.—Larger than the male, about half of each wing dull black, and the hind wings have two tails. The under side lacks the blue on the fore wings.

According to Morris, the larva is green, slightly pubescent, the head testaceous. On the back there is a small ray, and on the sides are nine oblique bands of obscure green. At the base of the feet is a marginal ray of yellowish green. The chrysalis is russety painted with brown. Food-plant, oak.

Gulf States, Illinois, Nevada, California, Arizona.

96. Thecla M Album, Bd.—Lec.

Expanse of wings 1.3 inches.

Upper surface rich glossy violet-blue, a broad outer border and costal margin of black. The hind wings have two tails, and a fulvous spot preceded by white at the anal angle.

Under side brownish gray, the fore wings with a single white line beyond the middle, bent inward on the second median venule, and then outward again below. This is continued across the hind wings, making a rude W before the tails, from this extending inward to the inner margin. Between this and the margin is a second line, the anterior half consisting of a series of shallow crescents, the whole edged outwardly with black, broken before the first tail by an orange patch which extends inward to the first line. Outer margin of wing white, with a large pale blue patch before second tail, separated from the anal angle by a black spot, a white spot in the subcostal area of the hind wings towards the base. Tails black, white-tipped. Lower part of knob of antennæ and costa fulvous. The female has the black border on the upper side broader than the male.

The larva is slightly pubescent, pale green, a little yellowish, with a dorsal stripe and seven oblique streaks of dark green. Head black, a marginal ray of yellow, slightly shaded with obscure green on its upper side. Food-plant, oak.

The chrysalis is brownish gray, with the anterior part and the wing-cases pale greenish gray.

Gulf States, Virginia, Pennsylvania, Ohio.

97. Thecla Favonius, Sm.—Abb.

Expanse of wings from 1 to 1.3 inches.

Upper surface brownish black, the male a little the brighter. Males with a blackish sex-mark in the cell of the fore wings, and a fulvous patch beyond the middle of the hind wings, and a smaller spot at the anal angle. The females have a similar patch, more or less distinct, on the fore wings; and both sexes have the anal portion of the outer margin with a white line just within the edge, continuous as a dull streak to the apex.

The under side is brownish gray, crossed by two white broken lines common to both wings, the inner forming a W on the hind wings similar to that on the under side of *M. Album*, the two converging from before back, the outer touching the points of the W, from which the two

run parallel to the inner margin.   The inner line is
edged with black within, and the outer black on the outside.   A broad fulvous band extends on the outside of
the second line from the outer point of the W to near the
apex, where it tapers to a point, with more or less black
between this and the outer edge.   In front of the tails
pale blue, sprinkled with black, with black at the anal
angle.   Subcostal patch as in *M Album*.

The larva is said to be yellowish green, with a dorsal
line and eight oblique streaks of dark green.   Marginal
line green, yellow below.

Chrysalis ash-gray, with two rows of blackish points
on each side of the abdominal rings.

Gulf States, South Carolina.

98. THECLA AUTOLYCUS, Edw.

Expanse of wings from 1.05 to 1.1 inches.

Upper surface brown, the fore wings with a large
patch of clear ochraceous in the outer half, in some
examples almost a broad band nearly from margin to
margin, or it may be subtriangular, with a blackish
sex-mark in the cell.   Hind wings with a patch of the
same before the tails, and a little fulvous at the anal
angle.   Tails two; between their bases there is usually
a black spot.

Under side brownish gray or fawn color, with two
white, more or less broken lines; the two on the fore
wings, and the inner one on the hind wings, edged with
brown, the other with a black border, the position of the
edging as in *Favonius*.   The inner line of the hind wings
makes a shallow W, the outer line sending out three dentations, one opposite each tail and one opposite the anal

angle. The outer line has a series of fulvous crescents outside the black, and beyond this a black spot at the anal angle, one between the bases of the tails, and between these there is a pale blue patch.

Texas, Missouri.

99. THECLA HUMULI, Harr.

Expanse of wings from 1.05 to 1.2 inches.

Upper surface blackish, with a slight blue-gray tint, the males with a blackish patch at the end of the cell of the fore wings. Along the outer margin of the hind wings is a series of more or less distinct pale blue spots, interrupted by a large orange crescent enclosing a black spot, the blue spot towards the anal angle with a similar black spot; an orange spot at the anal angle. Margin of hind wing black, with a white line within; two slender tails, tipped with white.

Under side gray, two blackish-brown lines crossing both wings, the inner edged externally, and very slightly internally, with white, and the outer on the hind wings faintly edged on the inside with white; these two lines converging a little posteriorly, the inner, on the hind wings, forming a very shallow W. The orange and black spots of the upper surface are repeated, but the . orange is enlarged to a patch; the anal angle is black, with an orange spot before it. Both wings have black outer margins, supplemented with white on the hind wings.

The larvæ are "green, downy caterpillars," feeding on the common hop-vine. The butterfly is to be seen in May.

Atlantic States, Mississippi Valley, Montana.

100. Thecla Acadica, Edw.

Expanse of wings 1.2 inches.

Upper surface dark grayish brown, costal edge of both wings red or reddish, the males with the usual subcostal sex-mark. Hind wings with one tail, a very slight angle in place of the second. The anal portion of the hind wings edged with white, and before this a distinct fulvous band.

Under side gray, varying from brownish gray to gray-white. At the end of each cell a short bar edged with white. Beyond this a bent row of black spots surrounded with white, those on the hind wings not reaching the inner margin. Beyond these is a row of black crescent-like spots, bordered within with white and without by fulvous, the latter fading out towards the apex of the fore wings; the last and the third from the last on the hind wings large, with the usual blue patch between them, these two with a black outer edging.

Of the preparatory stages of this species Mr. Wm. Saunders gives substantially the following. Found feeding on willow four succeeding years. Length .63 of an inch, onisciform. Head very small, pale brown and shining, drawn within joint 2 when at rest. Body above green, of a moderately dark shade, thickly covered with very short whitish hairs, scarcely visible without a lens. From joint 3 to 10 a dorsal line of darker green than the ground color. Dorsal region flat; rather wide; border, a raised whitish-yellow line, beginning at joint 3 and fading out on joints 12 and 13. Sides of body inclined at an almost acute angle and faintly striped with oblique greenish-yellow lines. A whitish-yellow line borders the under surface, beginning on joint 2 and extending round the hind end of the body. This line is raised the same as that bordering the dorsal ridge.

The chrysalis is .32 of an inch long, thickly covered with minute hairs. Color pale brown, with many dots and patches of darker shade, a dark ventral stripe from joint 7 to the end of the body. Sides with four or five short lines of dark brown.

Northern States, Montana, Nevada, Arizona.

101. THECLA EDWARDSII, Saund.

Expanse of wings 1.1 inches.

Upper surface pale wood-brown, the male with the usual subcostal sex-mark, hind wings with one short tail and an angle in place of the second tail; two faint blackish spots on the hind wings, one between the tail and the angle, and the other towards the anal angle, faint orange crescents before each.

Under side paler than the upper, two rows of spots across each wing as in the lines of the preceding species, but here they are shorter, with spaces between. The spots of the inner row, except the last two on the hind wings, are oblong and oval, each surrounded with white, the last two longer than the others. The outer row is a series of blackish crescents, edged on the inner side with white, on the outside with orange, fading out towards the apex of the fore wings, more prominent at the anal portion of the hind wings; the usual blue patch between the next to the last and the margin, and the two black spots of the other species. At the end of the discal cell a spot similar to the spots of the inner row.

Maine to Nebraska, Colorado.

102. THECLA WITTFELDII, Edw.

Expanse of wings from 1.5 to 1.7 inches.

Male.—Upper surface black-brown; fore wings have
a large oval stigma; hind wings have the edge of the
hind or outer margin on posterior half pure metallic
blue; a large fulvous spot in second median interspace
over a black spot on the margin; two tails, the posterior
one very long, measuring .24 of an inch on the anterior
side, the other .1 of an inch; black tipped with white.
Fringes of fore wings fuscous, of hind wings the same
to the upper median venule, then white, and next anal
angle long, brown, with a whitish line running through
them.

Under side dark brown, the outer margins narrowly
edged with white; the costal edge of the fore wings
next the base red. Both wings are crossed by two
macular white lines, the outer one submarginal, nearly
parallel to the margin, and quite regular, broken at the
venules, crenated on posterior half of hind wings, and
ending in an oblique streak up the inner margin; each
spot edged with black on outer side, and on fore wings,
in the median interspaces, there is more or less fulvous
outside the black. On the hind wings is a large spot
on the margin below the lower median venule, made
by blue-white scales on the brown ground; in the next
interspace are three deep red fulvous spots, diminishing
gradually in size, the outer one sometimes obsolete, the
largest with a black patch on its marginal side. Anal
angle black, overlaid on inner margin by white and a red
streak. The inner of the two lines is extra-discal, somewhat irregular, especially on the hind wings, and joins

the other at the lower median venule of the hind wings, then makes an angle in submedian interspace, and ends in a streak up the inner margin. In the cell of each wing are two parallel abbreviated white streaks or bars.

Female.—Upper side as in the male except the stigma. The tails measure .26 and .12 of an inch respectively. Under side as in the male.

This species differs from *Calanus* in the longer tails, in the larger size, and in the inner of the two lines on the under side being single, whereas in *Calanus* it is double or chain-like.

Indian River, Florida.

103. THECLA CALANUS, Hüb.

Expanse of wings 1.1 inches.

Upper surface dark wood-brown, with slight brassy green reflection, the males with the usual sex-mark; hind wings with two tails, both rather short; a fine, white marginal line from the anal angle to the longest tail, with a black mark between the tails.

Under side grayish brown; both wings crossed by two darker, broken lines, on the hind wings both lines white,

FIG. 73.

Thecla Calanus, male, the under side at the left.

edged on both sides, but more distinctly so on the outside of the inner line and on the inside of the outer line. These parts of the lines on the fore wings are distinctly white, but only now and then a few white scales on the other side of these lines. On the hind wings the outer

line shows the open W mark, but the inner is broken at
the second median venule, both extending some distance
up the inner margin. The black spot between the tails
is repeated, another at the anal angle, the space between
these spots, the line and the edge of the wing, being filled
with blue and black scales. Next to the outer line are
three orange crescents, one before each black spot, and
one in the first median interspace. At the end of the
cell on both wings is a short double bar edged on both
sides with white.

Var. LORATA, Gr.—Rob.—This is a form in which
the under side has an extra basal streak common to
both wings, composed of powdery dark blue scales.
This is slightly irregular, and is lost inferiorly among
the longer scales which clothe the inner margin of the
hind wings.

The larva, according to Mr. Wm. Saunders, has the
head small, pale greenish yellow, with a minute black
dot on each side. Body yellowish green, streaked above
with yellowish white, and thickly covered with fine,
short, white hairs; second joint of a darker shade of
green than the rest of the body. A dark green dorsal
stripe on joints 3 to 5, the full width of the dorsal crest;
narrow on the four terminal joints, almost obsolete on
those intermediate. A faint whitish dorsal line runs
through the entire stripe. Dorsal crest edged with yellowish white, most apparent where it borders the darker
portions of the dorsal stripe. Sides of the body with a
few faint oblique lines of yellowish white. Body margined on each side with the same color close to the under
surface, extending round the posterior joint.

After the last moult the body becomes more whitish

green, with the dorsal stripe greenish brown. On joint 5 a streak of dark brown crosses the end of the dorsal stripe, extending down the sides; with several brown dots along the sides. Joints 10 and 11 with an oblique brown streak on each side. The sides of the body with five or six oblique white lines.

Before pupating the larva assumes a delicate pink color.

The chrysalis is .4 of an inch long, pale brown, sprinkled with many dots of a darker shade, is thickly covered with short yellow hairs, and has a ventral line of dark brown.

The larvæ feed on oak.

Atlantic and Western States, Texas, Colorado, New Mexico. Var. Lorata is found in West Virginia.

104. Thecla Ontario, Edw.

Expanse of wings 1.1 inches.

Upper surface dark grayish brown, fore wings of male with an oval sex-mark near the costal margin; and one tail, and an angle in place of the second. From the base of tail to the anal angle a fine white marginal line, with a few fulvous scales at the anal angle. Between the first and second median venules is an indistinct marginal dark spot, in front of which is a small fulvous crescent.

Under side uniform light brown, with two transverse lines, common to both wings, much as in *T. Humuli.* The inner line of the fore wings is edged without with white, as is also the corresponding one on the hind wings. The outer line on the fore wings is indistinct at the extremities, and shows a very little external white edging through the middle; but on the hind wings this line is

distinctly white-edged internally. On the hind wings
the inner line forms a shallow W, the outer line being
dentate in this part of its course, both extending part
way up the inner margin. The black spot of the upper
side is repeated, with another at the anal angle. The
outer line has along its course from in front of the black
spot back to the inner margin a fulvous stripe, with a
large light blue patch between the black spots.

The body above is fuscous; beneath, abdomen ashy
brown, thorax blue-gray; palpi white tipped with black;
antennæ annulated white and black; club black tipped
with ferruginous.

New England States.

105. THECLA STRIGOSA, Harr.

Expanse of wings from 1.1 to 1.2 inches.

Upper side dark brown, the males without spots; but
the females sometimes have a fulvous spot near the anal
angle, and they are paler in color.

Fig. 74.

Thecla Strigosa, the lower
showing under side.

Under side pale reddish brown.
The outer part of both wings is
crossed by four irregular, rather
wavy, white lines, varying a little
in different specimens, but the
two inner ones on the fore wings
approach each other on the hind
margin. The third is shorter
than the second, and the fourth
reaches only to the middle of the
wing. The inner line on the hind
wings extends nearly across, then,
bending, runs part way up the inner margin, preceded in

the last part of its course by another line nearly parallel to it. Above the termination of these two is a circlet of white on the margin. The outer line is short, and limited to the middle of the wing. The fore wings have a submarginal row of indistinct brown lunules, edged on the inside with white; the hind wings have a similar series, which are bright red towards the anal angle, and edged on the inner side with black followed by white, and enclosing, next to the anal angle, a large black space which is nearly covered with blue scales. Beyond this is a small black spot; and there is another at the angle, surmounted by a red stripe, edged like the lunules, and extending up the inner margin. The lunules next the apex usually exhibit a few scales of red. The margins of both wings are edged with a fine whitish line.

The body is fuscous, beneath grayish white; palpi white, the upper joint black tipped with white; antennæ annulated black and white; club fuscous tipped with white.

According to Mr. Wm. Saunders, the larva feeds on a species of thorn (*Cratægus*). The following is his description of it:

Length half an inch. Head greenish brown. Body flattened, sloping abruptly at the sides. Color velvet-green, with a darker-colored dorsal stripe. The anterior edge of second segment yellowish brown, with a few darker dots; the middle segment is laterally striped with two or three faint yellow oblique lines, the last two segments have each a lateral yellow patch, and there is a faint yellow basal line from the fifth to the terminal segment. Under surface bluish green.

The chrysalis is .37 of an inch long, nearly oval; the

head-case rounded. The body is dark reddish brown,
with black markings, and thickly covered with fine
hairs. The anterior segments have many black patches
on their, and there is a dark ventral line from joint
6 to joint 12.

Atlantic States, Mississippi Valley, Montana, Colorado.

106. Thecla Smilacis, Bd.—Lec.

Expanse of wings 1.1 inches.

The upper surface dark brown; the middle area of
both wings, except a broad costal border, ochraceous
between the veins. Hind wings with two slender tails,
black tipped with white.

Under side yellowish green, sprinkled with brown
scales, the end of the cell and along hind margin of fore
wings, and the middle portion of the outer fourth of hind
wings, washed with brown having a violet tinge. On
the hind wings a curving white bar at the end of the cell,
and a straight subcostal bar shaded outside with ferruginous brown. Beyond the cell a prominent white line,
sinuous on the hind wings, heavily shaded internally
with ferruginous brown. Edge of wings white; on the
hind wings between this and the white line, from the
inner margin towards the apex, is a series of five brown-
black spots slightly edged internally with white, the first
and third supplemented by two others nearer the margin.
Between these spots and the edge the space is covered
with a mixture of white, black, and brown scales. End
of veins black.

It is said that the larva of this species is green, with
the head and feet blackish. It has four rows of red spots,

of which the two dorsal are formed of smaller spots, and the one on each side is composed of spots which are somewhat larger.

The chrysalis is grayish brown, with the abdomen more clear and reddish.

The larva feeds on Smilax.

Atlantic and Western States, Northwest Texas.

107. THECLA ACIS, Drury.

Expanse of wings about 1 inch.

Upper side of fore wings entirely dark brown, without any marks or spots. Hind wings the same color; each wing with two tails, the one near the anal angle much longer than the other. Close above this tail is a bright red spot, edged outwardly with black, and another at the anal angle. Fringes white.

Under side of both wings lead color. A very narrow black and white line crosses the fore wings, parallel to the outer margin; and an indented, irregular line crosses the hind wings, beginning near the middle of the anterior edge, and meeting just below the extremity of the body. Two long reddish spots are visible on this side, below which are four black ones.

The above is substantially Drury's description. A female from Florida Keys differed a little, as the following will show.

Color of the upper side brown, with the inner half of the hind margin of the fore wings and the inner half of the hind wings bright blue. Only the anal red spot is visible.

On the under side the common black and white line is prominent on both wings; outside this on the hind wings

23*

is a trace of another from the apex, meeting the inner
line before the shorter tail, but renewed again before the
inner or longer tail, from which it goes to the inner
margin. Before the space between the base of the two
tails and the inner line is a patch of orange shading
into yellow outwardly, with a little yellow outside the
outer line near the inner margin. Between the orange
patch and the outer margin is a spot of red and black at
the anal angle, with a patch of blue and black between.
Between the orange patch and the apex the outer line is
bordered externally by a black shading, and outside of
this some pale gray scales.

Basal third of costa orange. Antennæ black annulate with white; knob black tipped with orange.

Key West, Florida.

108. THECLA POEAS, Hüb.

Expanse of wings from .9 inch to 1.1 inches.

Upper surface blackish brown, in some specimens
entirely so, in others a few blue scales along the inner
margin of the hind wings, in still others nearly all the
hind margin blue, with the same color extending over the
base of the fore wings. In such blue-washed specimens
there is a series of marginal black lunules along the
outer margin of the hind wings. The hind wings have
two unequal tails, black tipped with white. The costa
of the fore wings red-edged, more distinctly seen on the
under side.

Under side brownish or russety gray, with two lines
beyond the middle, common to both wings, as in allied
species. The inner is clear white, edged on the inside with
narrow black, and farther in heavily shaded with reddish

orange; the line making an open W near the anal angle of the hind wings. Outer line black, more or less obsolete on the fore wings, on the hind wings shaded more or less with white on both sides; the usual black spot between the base of the tails and at the anal angle, with the blue patch between. In some examples the black extends as a shade along the margin towards the apex; in one specimen the black spot between the tails is preceded by a reddish-orange crescent. Both wings have traces of discal bars.

Southern States, West Virginia, Kentucky, Indiana.

109. Thecla Columella, Fab.

Expanse of wings 1 inch.

Male.—Upper surface grayish brown, a large, quadrate, blackish stigma in the end of the cell of the fore wings. The hind wings have one tail, and about two black spots near the anal angle.

Under side gray, with a distinct lilac reflection. The fore wings have a dark bar across the end of the cell, somewhat obscure; beyond the cell is a slightly-bent row of black spots, not reaching the hind margin, each convex outwardly and bordered with white. Outside this are two rows, parallel with the outer margin, of faint white lunules, with a dusky shade between the rows and outside the outer row.

The hind wings have the bar across the end of the cell, a dot in the cell and one above near the edge, the curved row of black and white spots continuous with the fore wings; all having a slight sprinkling of orange scales on the black. Outside the line of spots are also the two rows of white lunules and dusky shades, a little more distinct than on the fore wings. In the outer row, between the second and third median venules, is an orange patch with a black spot outside, the two colors blending on their edges; a black spot at the anal angle sprinkled a little with orange, and a blue patch between them. Fringe whitish, tail black.

Female.—Of the same color above as the male, except that the inner half of the hind wings is washed with pale blue, and the outer margin of the hind wings has about five black spots. The stigma of the fore wings absent.

The under side is the same as the male, but the spots and marks are more distinct. Body bluish gray above, whitish beneath.

Florida, Texas; occasional in New York.

110. Thecla Augustus, Kirby.

Expanse of wings from .9 to 1 inch.

Upper surface dark brown, tinted with rusty brown on the outer part of the hind wings, on some examples a slight greenish reflection.

Under side of the fore wings lighter than above and more distinct brown, with a faint transverse line marking the outer fourth, beyond which is a row of small dots continued across the hind wings. The basal half of the hind wings is quite dark brown, but the outer half is about the shade of the fore wings, the whole sprinkled with light purplish scales. There are no tails.

In Maine this butterfly is on the wing from the middle of May to the middle of June. The early stages and food-plant are unknown.

Eastern States.

111. THECLA IRUS, Godt.

Expanse of wings from 1 inch to 1.25 inches.

Three forms have until recently been included in this species, two of which are still retained as varieties, while the third is set apart as the next species.

Var. ARSACE, Bd.—Lec.—This has the upper side of the wings dark brown, with greenish reflections. The stigma on the fore wings of the males is three times as long as wide.

Under side of the wings blackish brown on the basal half, and lighter beyond. The outer part of the fore wing is overlaid more or less with white scales. The tooth of the hind wing next outside the anal angle curves outward, and there is a more or less prominent black spot within the margin between the base of the tooth mentioned and the one next outside. In the hoary outer space of both wings is a transverse row of brown points, clouded somewhat on the hind wings.

This variety is found in the Atlantic and Western States.

Var. MASSII, H. Edw.—This form has been found only on Vancouver Island.

Morris says the larva of this species feeds on Vaccinium, and is yellowish green, with two dorsal interrupted lines; one lateral line and eight oblique streaks pale green.

112. THECLA HENRICI, Gr.—Rob.

Expanse of wings from .85 inch to 1.1 inches.

This species closely resembles var. *Arsace* of the preceding species in form and color as well as in markings,

s

but differs in having the outer part of the hind wings
somewhat rust-colored, the teeth shorter, and the first one
beyond the anal angle not curving outward, and in wanting the black spots on the under side of the hind wings
between the bases of the first two teeth. The stigma on
the fore wings of the males is shorter and wider, being
only twice as long as wide. It averages smaller than
var. *Arsace.*

According to Mr. Edwards, the eggs are deposited on
wild plums at the base of the flower-stalks. When
the young larvæ hatch from the eggs they ascend the
flower-stalks and eat the newly-forming fruit. A hole
is eaten into the fruit large enough for the head to
enter, and thenceforward the caterpillar spends most of
his time with his head in the cavity, growing with the
growth of the plum, until the whole interior is eaten
out.

The eggs are shaped like those of *Lycæna Violacea,*
flattened and depressed : about this depression the ridges
are reduced, and the surface is finely reticulated, but
elsewhere there is a white frosting of peaks and ridges.
Color whitish green. The lace-work seems to be separable : in one case where the egg had been rubbed by
a leaf, apparently, a film was turned up, and the real
surface was seen to be delicate green.

In from five to six days the larva hatches. Length
.04 of an inch, oval, broadest anteriorly, the base
flattened ; dorsum high, sloping posteriorly ; summit a
little flatteued, with two rows of long, recurved white
hairs on each side. Color brownish yellow.

Five days afterwards it moults the first time, when it is
.08 of an inch long, with a red-brown dorsal stripe stop-

ping at joint 12; on each side of this dull yellow-green, with a macular brown subdorsal line. Sides sloping, a little incurved, red-brown, with a broken yellowish line; a similar but continuous line along the basal ridge. Body much covered with short, stiff, brown hairs. Head cordate, smooth, yellow-green.

It moults again in three or four days, when it is .12 of an inch long, shaped much as before, with a tuberculous ridge across each segment from 4 to 11. Color red-brown and dull yellow-green. Dorsal band red, tapering posteriorly to a point, with a central green line; outside this band a green one, with a red-brown speck on each segment. Sides red-brown, with green central line; basal ridge green.

In five days it moults the third time, and measures .3 of an inch; but seven days later, when ready to pupate, it is .56 of an inch long. The summit of the dorsum is flattened, a little concave, covered by a broad band cut by a paler line; the remainder of the elevated ridges yellow-green. Sides red-brown, with an indistinct green line. Basal ridge green, head yellow-green. One day after moult the color changes to port-wine red; the subdorsal area remains yellow-green, but is red-tinted on the posterior part of each segment; the sides the same red; a pale red line along the basal ridge; about the spiracles a little green.

Chrysalis .3 of an inch long; black or blackish brown, with obscure red band, and on each side a narrow black stripe in the middle of the abdomen. In this stage the species hibernates.

West Virginia, Maine.

113. Thecla Niphon, Hüb.

Expanse of wings from 1 inch to 1.15 inches.

Upper side dark blackish brown, with a large rusty brown space in the middle of each wing of the female, but only near the anal angle of the male; the males with a small, oval, subcostal sex-mark.

Under side light brown, sprinkled with white scales so as to be hoary, especially in a band beyond the common white line. Fore wings with a brown bar at the end of the cell, edged outside with white; a brown spot farther towards the base. Beyond the cell a somewhat zigzag white line, not reaching the hind margin, bordered within by a broader brown line. Beyond this a submarginal row of dark brown lunate spots, shaded outside with white, and in some examples sending white rays to the white spots in the fringe. The hind wings with the basal third quite hoary, defined externally by dark brown. Cutting the outer third is a tortuous white line shaded within with dark brown, the middle third of wing but little hoary. The submarginal row of lunules continuous, shaded a little with white externally, the space between this and the white line quite hoary; a little washing of white along the outer margin. There are no tails, but about three prominent teeth or angles to each hind wing.

The larva is green and pubescent, longitudinal stripes on the back, the middle one pale yellow, the other two white. Near the feet there is usually a small marginal white line. Head brown.

The chrysalis is grayish, with four rows of small spots, of which the two middle ones are blackish and indistinct, the others ferruginous.

Food-plant, pine.

Atlantic and Western States, Colorado.

114. THECLA LÆTA, Edw.

Expanse of wings from .9 inch to 1.1 inches.

Male.—Upper side black, near the base of the fore wings a few scales of deep metallic blue; next the anal angle a band of the same color, which extends half-way along the outer margin, sprinkled with black. Beyond this a fine line of blue scales follows the margin to the outer angle. Anal angle edged with red; fringe gray.

Under side of hind wings and apex and costal margin of fore wings slate-blue, with a green reflection; costal edge of fore wings red; disk of same wings smoke-color. Beyond the cell, on the costal margin, a transverse, abbreviated series of five small red spots, edged outwardly with white, the last two obscured by the smoky hue of the disk.

The hind wings have two series of red spots parallel to the outer margin, those of the exterior small, and towards the outer angle minute, each more or less surrounded by a delicate white border, in which are a few black scales; the inner series crosses the middle of the wing, is sinuous, the spots large, brighter red and crescent-shaped, bordered outwardly with white in which are a few black scales. Edge of wing at anal angle and at the intersection of the adjoining venules red. Body black above, beneath white. Antennæ annulate white and black; club black, red at tip; palpi white.

Female.—Upper side black, the base of fore wings and hind margin for two-thirds its length, and all of hind wings except the costa, dark metallic blue. Under side greenish gray, losing the green tinge on the hind margin of fore wings. In addition to the five spots on disk of male there are two blackish, rather indistinct spots, below the others, nearer the base.

Larva and food-plant unknown.

Maine to West Virginia; Atlantic City, N.J.; Arizona.

115. THECLA TITUS, Fab.

Expanse of wings from 1.2 to 1.5 inches.

Upper surface dark wood-brown, the hind wings with a marginal row of seven orange spots from the anal angle towards the apex or outer angle. On some specimens these are partially or wholly wanting.

Under side grayish brown, with lilac reflections. The fore wings with two transverse rows of black spots, the inner edged on the outside with white; the outer smaller, touched without with vermilion, and slightly with white on the inside. Hind wings with two similar rows, but the outer one more distinct and with prominent vermilion spots on the outside, the two anal spots blended, each vermilion spot with a few black scales next the margin, and some white at the anal angle. Cell with two dashes at the end, placed end to end.

According to Mr. Saunders, this species is to be found on wild cherry and plum. When full grown it is .7 of an inch long, elliptical. Head very small, bilobed, black, and shining, with a streak of dull white across the front above the mandibles. Body above dull green, with a yellowish tint, especially on the anterior joints, and thickly covered with very short brown hairs. These arise from small pale yellow dots, which are slightly raised. A dark

green dorsal line from joint 2 to joint 4 ; a patch of dull
pink or rose color on anterior joints, faint on joint 2,
covering but a small portion of its upper surface, nearly
covering the dorsal crest on joint 3, and reduced again
to a small faint patch on joint 4. On the posterior joints
is a much larger rosy patch, extending from the posterior
of joint 9 to the end of the body ; joint 9 merely tinged,
enlarged on joint 10 to a considerable-sized patch widening posteriorly ; behind this the body covered with rosy
red. The side of joint 10 close to the under surface
has a streak of the same color, and there is a faint con-
tinuation of this on joint 9. There is a wide dorsal
crest from joint 3 to joint 9. Posterior part of body
suddenly flattened, sides acutely sloped.

The chrysalis is .45 of an inch long, glossy pale brown,
with many small dark brown or blackish dots, and thickly
covered with very short brown hairs, too fine to be seen
without a lens.

New England to Arizona.

SUBFAMILY LYCÆNINÆ.

In this the palpi project in front more than the length
of the head, and the antennæ do not reach to the middle
of the fore wings. The anal angle is rounded, with or
without one or two slender tails on the outer margin of
the hind wings.

116. FENISECA TARQUINIUS, Fab.

Expanse of wings from 1.25 to 1.4 inches.

Upper surface yellow, somewhat orange-tinted. The
fore wings have a dark brown border, irregular on the

inside, and narrow on the hind margin; it sends an angle inward near the apex, a bar is at the end of the cell, and there is a separate spot below, and a black basal dash. The hind wings have the anterior half dark brown, the lower edge crenate, with a few small spots along the outer margin.

Under side of fore wings pale yellow; the costal and terminal margins and the whole of the hind wings reddish yellow washed with white, with numerous white circles whose enclosed spaces lack the white. The spots in the pale yellow part are dark brown.

The larva is said to feed on thorn, alder, wild currant, Viburnum, and Vaccinium: it is green, with three dorsal white stripes, and one at the base of the feet.

Chrysalis grayish; back darker, marked with prominent tubercles.

There seem to be two broods of this species in a season, as the butterflies have been taken in the same locality in June and again in August.

Atlantic States, Mississippi Valley.

117. CHRYSOPHANUS DIONE, Scud.

Expanse of wings from 1.5 to 1.8 inches.

Upper surface brownish gray, with slight coppery reflections. Hind wings without tails, but angled, the anal angle a little produced. Hind wings of female with five black lunules along the margin from the anal angle, preceded by continuous orange crescents, the orange extending to the posterior angle of the fore wings. Some of the heavy marks of the under side show through. The males have the black lunules, but with very little orange.

Under side grayish white, both wings with a double marginal row of dark spots, the outer obscure on the fore wings and the apical portion of the hind wings. Between the two on the anal half of the hind wings, and a little at the posterior angle of the fore wings, the space is filled with orange, more prominent in the female. The fore wings have a sinuous row of elongate, bright black spots across the disk, with two similar spots in the cell, a bar at the end, and a spot below. The hind wings have spots in the same position, with two subcostal and one near the inner margin.

Iowa to Kansas, Nebraska.

118. Chrysophanus Thoe, Bd.—Lec.

Expanse of wings from 1.4 to 1.5 inches.

Upper surface of the male coppery brown, with violet reflections, a narrow terminal blackish border. The hind wings have an orange fulvous terminal border, crenate within, and containing five black lunules on the black edge. The female has the fore wings fulvous, with a broad blackish border, and the black spots of the under side, except the terminal double row; hind wings as in the male, but more blackish.

- Under side of fore wings fulvous, the terminal border of hind wings orange fulvous; under side of hind wings and terminal border of fore wings grayish white, the latter growing narrow from the apex back. The position of the black marks is the same as in *C. Dione*, but they are heavier.

According to Mr. Saunders, the eggs are nearly round, a little flattened at the apex, flattened also at the base. Color greenish white, thickly indented; at the apex

24*

is a considerable depression, around which the indentations are small, increasing in size as they approach the base.

The food-plant is dock, *Rumex crispus*, and there are two broods in a season.

Maine to Nebraska, Ohio, Kansas, Colorado.

119. CHRYSOPHANUS EPIXANTHE, Bd.—Lec.

Expanse of wings from .85 to 1 inch.

Upper surface of the male brown, with a strong violet reflection, the female more grayish brown, with little of the violet reflection. The spots of the under side, except the subterminal row, show through somewhat; and there is a sinuous orange fulvous line from the anal angle of the hind wings along the margin, fading out about the middle.

Under side yellowish gray, the hind wings of the male but little yellow-tinted. The orange fulvous line is repeated, only more distinctly, with scarcely a trace of a subterminal row of black spots on the hind wings. On the fore wings are three large spots of this row from the posterior angle towards the apex. The discal row of spots on the fore wings is distinct, as are also the two spots and bar of the cell, and the one below the cell. On the hind wings the spots are all small, with the bar across the cell absent, as also one subcostal spot.

According to Mr. Saunders, the eggs are nearly round, slightly flattened at the apex, flattened also at the base. Color milk-white, thickly indented; a deep depression at the apex, and around this a number of indentations, which are nearly uniform in size all the way to the base,—in this respect differing from the eggs of *Thoe.*

Food-plant unknown, but probably dock. The butterfly flies through the last of June and early part of July. Maine, New England, Kansas.

120. CHRYSOPHANUS HYPOPHLEAS, Bd.

Expanse of wings from .9 to 1 inch.

Upper surface of fore wings shining fulvous, with a blackish brown border, narrow on the costa and hind margin. Beyond the cell is a row of prominent black spots, the first three and the other four in sets nearly confluent; a spot and bar in the cell. Hind wings blackish brown, with a terminal fulvous band, not reaching the apex, containing four black spots on the edge; at the end of the cell a black bar.

Under side of fore wings fulvous, paler than above, the apex and the whole of hind wing gray, the gray of fore wings narrowing backward to the posterior angle. The spots of the upper surface are repeated, with some of the terminal border of the fore wings. The hind wings have two subcostal black spots, two spots in the cell and a narrow bar at the end, and a spot below the cell, besides the usual curved and sinuous discal row, the spots of this row white outside. Along the outer margin is a crenate orange-red line, shaded within with faint reddish, the whole between two faint rows of blackish spots.

Aberr. var. FASCIATA, Strecker, has the transverse row of spots on the fore wings much enlarged, and blended together so as to make an irreg- ular band.

FIG. 75.

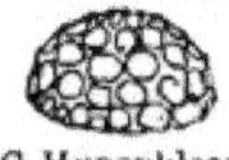

C. Hypophleas, egg, × 16.

The egg (Fig. 75) is shaped and marked like that of the other species, as shown in the figure. It is pale green, overlaid with a white net-work.

According to **Mr. Saunders**, the larva is elliptical, flattened on the under side, dull rosy red, with a diffused yellowish tint on the sides, most distinct along the middle joints. The dorsal line is a deeper shade of red. The body is downy with minute yellowish hairs. This description was taken from a larva not fully grown. Mr. S. H. Scudder says that the larvæ are provided with long hairs sweeping backward behind their bodies, most of them arranged in longitudinal series.

FIG. 76.

C. Hypophleas, larva.

The chrysalis is attached by the end of the abdomen, and is closely girt to the object to which it is attached, as in Fig. 77.

FIG. 77.

C. Hypophleas, pupa.

Northern, Middle, and Western States; California.

121. LYCÆNA LYGDAMUS, Doubl.

Expanse of wings 1 inch.

Upper surface silver-blue; the males with only the edge of the wings black; the females with a rather broad black terminal border, a little expanded on the apex, and extending round on the costa. Fringes long, fuscous.

Under side uniform gray-brown. The fore wings have a small round black spot in the cell, a bent bar at its extremity, edged with white. Across the disk is a curved row of large round black spots, the two lower ones connected, all annulate with white. The hind wings have a small black spot in the cell, another on the costal margin, a narrow stripe at the end of the cell, and a sinuous row of spots within the margin similar to those of the fore wings; all edged with white. The under side of the female is a little paler than that of the male.

The larva and food-plant are unknown. The butterflies appear in April.

Michigan, Wisconsin to Georgia, West Virginia.

122. LYCÆNA SCUDDERII, Edw.

Expanse of wings 1.1 inches.

The male, in size, form, and color, resembles *Ægon* of Europe. Upper side dark violet-blue, outer margin of both wings and costal margin of hind wings edged with black, costal margin of fore wings with a fine black border. Fringe white.

Under side dark gray. The fore wings have an oval black discal spot, and a transverse, tortuous series of six black spots, all edged with white, the one next the posterior angle double, the fifth twice as large as the others; on the outer margin a double series of faint spots.

Hind wings with four black spots near the base, one being very minute and close to the inner margin; a bar at the end of the cell, and a series of eight spots in a double unequal curve; all of which, as well as the basal spots, are edged with white. There is a marginal series of six or seven metallic spots, each surmounted with a spot of fulvous bordered inwardly by a dark crescent. These metallic spots are edged outwardly, and sometimes replaced, by black. Ends of venules expanded into small black spots.

A more common form of the female has the base of both wings violet-blue, and the black marginal spots of the hind wings distinct, the two or three next the anal angle surmounted with fulvous. Under side as in the male.

A variety of the female has the upper side brown, with a black discal spot on the fore wings. Hind wings with a marginal row of obsolete spots surmounted by grayish crescents.

Under side pale buff, fore wings as in the male, except that the discal spot is preceded by a small double spot, and all the spots are larger. On the hind wings the spots are less distinct, and some of them are wanting. The transverse series is set in a band of white; marginal spots without the metallic gloss.

New York, Michigan, Wisconsin.

123. Lycæna Pseudargiolus, Bd.—Lec.

Expanse of wings from .9 inch to 1.4 inches.

This species is subject to great variation. According to the recent observations of Mr. W. H. Edwards, published in his "Butterflies of North America," it may be known under the following polymorphic forms, most of which have heretofore been regarded as distinct species:

Winter form 1, Lucia, Kirby.

Winter form 2, Marginata, Edw.

Winter form 3, Violacea, Edw.

Dimorphic, male, Nigra, Edw.

Var. Cinerea, Edw.

Spring form, Pseudargiolus, Bd.—Lec.

Var. Arizoniensis, Edw.
Summer form, Neglecta, Edw.
Pacific form, Piasus, Bd.
Var. Echo, Edw.

The first of these, *Lucia*, coming from hibernated chrysalides, is to be found in spring in Alaska, British America to Quebec, Anticosti, New England, New York, and Colorado.

Marginata has a similar range, except that it does not occur quite so far north, being found in Ontario, Quebec to Long Island, and Colorado.

Violacea has a more extended range, being found in Alaska, vicinity of Lake Winnipeg, British America, Ontario, Quebec, Anticosti, New England to West Virginia, and Colorado. In the southern part of this territory a black male *Nigra* has been found.

Var. *Cinerea* appears to be the winter form in Arizona; under side ash-gray, with the markings obscure.

Pseudargiolus is a spring form ranging from Racine, Wisconsin, south to Tennessee, and on the east extending from Pennsylvania to Georgia.

Var. *Arizoniensis* is a small form of this, found in Arizona.

Neglecta is a summer form when there is more than one generation during a season, ranging from Canada through New England to West Virginia and Georgia; occurring also in Montana and Nevada.

Piasus and its variety *Echo* are Pacific slope forms, found in California and Arizona.

In general terms, the upper side of the wings of the male is a deep azure-blue, with a delicate terminal black line. Fringes black on the apical part of the fore wings, but white barred with black on the rest of the fore wings and on the hind wings.

The fore wings of the female have a broad blackish outer border, in some examples extending along the costa. The hind wings have a blackish costa, and a row of dark spots along the outer margin. The ground color is usually a lighter blue on the females than on the males.

The under side is a very pale silvery gray, with a silky lustre, and there are the following pale brown markings: a row of spots along the outer margin, each preceded by a crescent; a curved row of elongate spots across the disk of the fore wings; and several small spots on the basal part of the hind wings.

Fig. 78.

Lycæna Pseudargiolus, form Lucia, under side.

The form *Lucia* has the terminal spots of the under side so enlarged and run together as to form a terminal band, and the spots on the basal part of the hind wings are enlarged and run together so as to form a more or less complete triangular discal patch, as shown in Fig. 78. This and *Marginata* are the smaller forms.

Fig. 79.

Lycæna Pseudargiolus, form Violacea, under side.

Form *Marginata* has the terminal band of the under side as in *Lucia*, but the spots on the basal part of the hind wings do not coalesce.

Form *Violacea* has the dark points and crescents on the under side of the wings quite prominent, but they do not coalesce, either in the outer border or in the basal portion. The black male *Nigra* has the under surface the same as in the blue *Violacea*, but the upper surface is black.

Form *Pseudargiolus* is the largest of the series, expanding 1.4 inches. The upper surface of the male usually has a terminal border to the hind wings of the same shade of blue as the fore wings, the middle area of the hind wings a little paler than this border or the fore wings. The spots on the under side are much smaller than on any of the preceding forms.

Form *Neglecta* resembles *Pseudargiolus*, but is smaller, not expanding more than 1.1 inches; spots on the under side small, as on *Pseudargiolus.*

The Pacific form *Piasus* is of a deeper blue, the under side bluish, with the border spots subobsolete.

The winter forms of these butterflies deposit their eggs in the clusters of flower-buds of dogwood (Cornus), the young larvæ obtaining their first food by boring into the buds, but later eating their way into the ovaries. The next brood of eggs are deposited on the flower-stems of rattleweed (*Cimicifuga racemosa*), while the fall brood are to be found on *Actinomeris squarrosa*, and probably on *A. helianthoides.* In confinement they have been known to eat several other plants; and it is probable that where rattleweed does not grow, the second brood of larvæ feed upon something else.

The eggs are .02 of an inch in diameter, round, flat at base, the top flattened and depressed; the surface covered with a white lace-work, the meshes of which are mostly lozenge-shaped, with a short rounded process at each angle. Ground color delicate green.

In from four to eight days a larva .04 of an inch long hatches from the egg. This has the under side flat, legs retractile, upper side rounded, highest at joint 4, from which the dorsum and sides slope gradually to joint 13. Surface pubescent. On each side of the dorsal line is a row of white clubbed hairs, with similar hairs at the base and in front of joint 2, making a fringe round the body. Color greenish white or brownish yellow. Head minute, obovoid, retractile, black.

After the first moult, which occurs in from three to five days, the length is from .07 to .08 of an inch; the color of the spring and fall broods is brownish yellow, that of the summer brood the same, and also greenish white and reddish.

The second moult occurs in from three to five days more, the length being from .12 to .16 of an inch. The shape is as before, but the dorsum is covered from joint 3 to joint 10 by a low, broad, continuous, tuberculous ridge, cleft to the body at the juncture of the segments, the anterior edge of each joint depressed, the sides incurved. Joint 2 is more flattened than before, and the outer border is thickened into a rounded rim, leaving within the curve a flattened, depressed space. Color in spring pale green, the dorsum whitish, usually a reddish dorsal line from joint 3 to joint 10. In summer variable, buff or pale green without spots, joint 2 brown; in some the dorsum and sides are mottled with dark green and brown; occasionally one is wine-red, or red with a white basal stripe, and white along the tuberculation. In fall dull green, more or less marked with brown.

The third moult takes place in three or four days

more, but there is little change from the former period. The fourth or last moult occurs three or four days afterwards, and in five or six days from this the larva is ready to pupate.

The mature larva is .4 of an inch long in the spring and fall broods, and from .5 to .55 in the summer brood; the shape as in the preceding stages. The color is variable. In spring, usually the ridge is whitish, often stained red, or it is brown, light or dark; the upper part of the side olive-green, with a darker green or sometimes a dull red patch along the posterior edge of each joint. Below this area it is pale green, and along the base more or less brown. Joints 11 to 13 are mottled in shades of green, often with brown, and joint 2 is either green or brown; if the latter, then with a brown patch in the depression. Color in summer, sometimes white or delicate green, joint 2 being brown; or the ridge is light green and the sides dark, often with brown patches over all; or light green, with a dorsal macular deep green band, and a similar one along the base; or the whole surface may be wine-red or even chocolate-brown. Color in fall, green, with more or less brown in irregular patches. Head dark brown.

The chrysalis is dark brown or yellow-brown, varying, the wing-cases dark, and sometimes green-tinted. On the abdomen are two subdorsal rows of blackish dots, sometimes a dark dorsal line.

In addition to the food-plants already given, *Apios tuberosa, Erythrina herbacea, Spiræa salicifolia, Ceanothus Americanus,* Cornus, and Ilex may be specified.

124. Lycæna Comyntas, Godt.

Expanse of wings from .7 to 1 inch.

Hind wings with one thread-like tail. The males are dark violet-blue above, with a narrow blackish outer border. Along the outer margin of the hind wings are several black spots, and usually one or two orange crescents. The females are blackish brown, some specimens with bluish at the base, the black spots of the hind wings often obscure.

Fig. 81.

Lycæna Comyntas, left wings, under side.

Under side whitish gray; both wings with a double row of spots along the outer margin, the inner row crescents; between this and the cells a row of black spots circled with white, the row on the hind wings broken twice. There is also a bar at the end of each cell, and on the hind wings a spot in the cell, and one above. Near the anal angle are two or three orange crescents, enclosing each a black spot with a circle of silver scales.

According to Mr. Edwards, the eggs of this species are deposited on red clover and *Desmodium Marilandicum*. They are round, flattened, depressed at top, covered with a frost-work of interlaced points. Color delicate green.

The young larvæ are .05 of an inch long, cylindrical, of a yellowish color, except two rows of white points along the back, and one near the base on each side. A long, curved, white hair has its origin in each of these points. Head black and shining, retractile, nearly as large as joint 2.

After the first moult they are .08 of an inch long,

onisciform, flattened, the dorsum flat at the top and sloping towards the base. Color greenish, the whole surface irregularly dotted with black; and from most, but not all, of the dots come white hairs, those on the dorsum curved back, those nearer the base curved partly downward and partly back. Head obovate, long, and narrow, smaller in proportion to the second segment than at the last stage, and partly concealed, even when active, in joint 2. Color black.

After the second moult they are .12 of an inch long, broader and flatter than before. On each side of the narrow dorsal ridge is a slightly-raised ridge, caused by the tubercles; at the base of the body a fold, and the hairs from this and the ridge are longer than elsewhere. Color green, but with a rusty tint, caused by the numerous reddish points. Above the fold these take the form of a line or slight stripe.

Moulting again, the length is .2 of an inch, and the color clear apple-green; the crests of the dorsal ridge, and also the folds at the base, are whitish; part of the way a reddish line on each side of the fold; also a double oblique line of pale green on each side of each segment.

After the fourth moult they are .36 of an inch long, and the width is about one-fourth the length. They are of the same general form as before, but highest in front, and sloping backward. Color greenish, with darker green lines, and oblique vinous lines on the sides. Head black.

The chrysalis is .26 of an inch long, shaped much like the mature larva; of a greenish, brownish, or sordid white color, with three rows of black dots, and sparingly clothed with whitish hairs.

The above description was taken from larvæ fed on Desmodium. Those that fed on clover differed in coloration, as follows: color russet, varying towards vinous, interspersed with green; at third moult some were pale green dorsally, the white being caused by the tubercles, the sides vinous, the dorsal stripe and oblique lines vinous; others had the back as well as the sides vinous, and this variation and character persisted to maturity. The chrysalides of these larvæ were sordid white on the upper surface and lower side of the abdomen, the former specked with brown; dorsal stripe brown, as were also the dots; under side of thorax and of head-case, and the whole of the wing-cases, apple-green.

Atlantic States to the Rocky Mountains, Colorado.

125. LYCÆNA FILENUS, Poey.

Expanse of wings from .75 to .95 of an inch.

Upper side of the male blue, with a slight black border; fringes white. In both sexes there is a small round black spot near the outer edge of the hind wings. The female is blackish brown, with the middle area of the basal half of both wings washed with blue.

Under side ash-gray, usually paler in the male than in the female, with a discoidal crescent on the middle of each wing, bordered on each side with white; and three sinuous common bands, formed of small black spots circled with white, of which the outer are a little less distinct and somewhat sagittate. The space which separates the inner band from the median is usually whiter than the rest of the surface, and forms a band of small white quadrangular spots. The base of the hind wings has a transverse row of three very black points annulate with white, of which the external, out of line with the others, is the largest. The hind wings have on the outer edge and near the anal angle a black eye-spot, more or less annulate with yellow, sprinkled behind with golden-green atoms.

Gulf States.

126. LYCÆNA ISOPHTHALMA, Herr.-Schaeff.

Expanse of wings .75 of an inch.

Upper surface brown, slightly grayish at the base and along the costal margin, a row of five blackish spots along the outer margin of the hind wings, the three next the anal angle the most distinct, the other two sometimes almost obliterated.

Under side of nearly the same color as the upper, slightly hoary at the base, the wings crossed by about seven rows of elongate white spots and an inner row of white circles. There is a submarginal row of six conspicuous black spots on the hind wings. The spot nearest the anal angle is almost entirely covered with metallic green, and the rest of the spots present detached scales of this color, generally on the side nearest the margin.

Palpi dark brown above, white beneath, with a few black hairs; antennæ annulate black and white; club reddish brown, red at the tip.

Florida, Georgia.

127. LYCÆNA EXILIS, Bd.

Expanse of wings from .6 to .7 of an inch.

Male.—Upper side brown, sometimes reddish brown, bluish at base of both wings, and slightly fuscous along the hind margins; hind wings bordered by a series of

round fuscous spots; fringes long, pure white, except at
the outer angle of the fore wings and against the lower
median interspace, where they are fuscous.

Under side of fore wings dark gray at base, without
markings, fulvous on disk, crossed by interrupted white
streaks; the margin gray, presenting a series of obsolete
lunules, in front of each of which is a white border,
forming a broken line like those on the disk. Hind
wings dark gray at base, paler gray over part of the
disk, then brown, followed by a broad white submarginal
space; the base marked by three fuscous points placed
transversely, and the brown disk crossed by white
streaks as on the fore wings; outer margin bordered
by a row of black spots, of which the first, next the anal
angle, is duplex and covered with metallic green scales;
the next four are large, dead-black, and metallic only at
their base; the next two wholly covered with metallic
scales.

Female.—Same size, paler-colored above, marked like
the male.

This description is of specimens taken in Texas.
Boisduval's description of those taken in California is
as follows:

Upper side clear brown; hind wings paler, with a
blackish border.

Under side of fore wings very clear brown, with white
interrupted transverse striæ, more or less distinct.

Under side of hind wings white, with brown striæ,
and a marginal row of seven black ocelli powdered with
golden atoms.

California, Arizona, New Mexico, Colorado, Texas,
Florida.

128. Lycæna Ammon, Luc.

Expanse of wings 1.05 inches.

Male.—Upper surface clear violet-blue, the edge of the wings black, and a small black spot near the anal angle. Fringes white, cut with black at the ends of the veins, broadly so on the fore wings.

Under side dark gray. The fore wings have a gray bar at the end of the cell, with a white line on each side; beyond the cell a curved row of intervenular gray spots, each with its inner and outer border of white. Along the outer fourth of the wing is a broad white band, dentate without, and a subterminal row of white lunules enclosing gray spots, the anterior half of lunules somewhat dusky. Edge of wings black, with a white inner border.

Hind wings with the same markings, the white band broader; in addition, the basal half has three round black spots annulate with white,—one in the cell, two near the costal margin,—and a gray spot near the inner margin. Near the anal angle are two round black spots, with a more or less complete annulus of metallic blue scales, the outer spot having before it an orange lunule.

Female.—Upper surface the same as in the male, outer part of hind wings a little paler, with costal and outer border of black; the latter on the fore wings a little mottled with blue, and on the hind wings containing a series of blue lunules and two bright black spots, the outer, or one farthest from the anal angle, with an orange lunule before it. Under side the same as in the male.

Food-plant and larva unknown. The species is a native of Cuba, but has been found at Indian River and in Southern Florida.

129. Lycæna Theonus, Luc.

Expanse of wings from .9 to .95 of an inch.

Male.—Upper surface violet-blue, rather pale, except at the terminal border. The marks of the under side can be seen through the blue.

Under side white, crossed by seven or eight brown-gray stripes, appearing along the costa as though this were the color, and as if there were eight more or less wavy white transverse lines arranged in pairs enclosing a space of the ground color a little darker. Along the outer margin two rows of spots, the outer roundish or oval, the inner lunate. On the fore wings the second, fourth, fifth, and sixth lines do not reach more than half-way across the wing, leaving a large white space. On the hind wings near the anal angle are two round black spots in place of two of the gray, each containing a circle of metallic violet scales; the spots annulate with pale yellow. The lines on the hind wings are more broken up into spots than those on the fore wings.

Female.—Upper surface white, with a violet-blue tint, more prominent on the basal half, with a broad outer border extending round the costa to the base. On the hind wings this border contains a series of white lunules, the two next the anal angle enclosing a round black spot. The markings of the under side show through more plainly than in the males. Under side the same as in the males.

Palpi black; antennæ black and white; club black, tipped with white; body black above, white beneath, with a yellowish tinge.

Florida Keys.

FAMILY HESPERIDÆ.

This family may be known by their robust bodies and their triangular fore wings, and by the knob of the antennæ usually ending in an attenuated portion,—that is, mostly bent or hooked; they have six feet adapted for walking. In their robust bodies and coarse scales, which are not closely appressed, these butterflies resemble some of the higher moths. When in repose, the wings are either spread or closed back to back and thrown back so that the costal edge of the hind wings is next to the costal edge of the fore wings.

The family is divided into two sections, which bear to each other somewhat the relation of families.

SECTION I.

The butterflies in this division have the knob of the antennæ thick ovoid or elongate ovoid. The larvæ are more or less spindle-shaped. The pupæ are somewhat conical, like those of the moths, smooth, and found in puparia composed of leaves fastened together, in which the larvæ seclude themselves.

130. CARTEROCEPHALUS MANDAN, Edw.

Expanse of wings 1.12 inches.

Upper side of wings dark brown, overlaid with a few yellowish scales, and marked with dull ochraceous spots, as follows: one row extends along the outer margin, often nearly obliterated; another across the disk, or between the margin and the end of the cell, two of which are out of line and nearer the outer margin; the cell is

more or less filled with the yellow, mostly in the form
of two patches, and there is a small spot resting on the
lower side of the cell near the middle, and another below
the submedian vein, about one-fourth of the distance from
the base. The hind wings have a row of small spots
along the outer margin, a row of larger spots within
this, across the disk, and one near the base of the wing.

Under side of fore wings much paler than the upper,
the light markings much larger and coalescing. The
hind wings are of the same yellowish as the fore wings,
the spots of the upper side repeated, but larger and of
a white color, ringed with brownish, with an additional
spot above the basal. The veins are all brownish.

White Mountains, N.H., and Maine, where it flies
about the middle of June.

131. Carterocephalus Omaha, Edw.

Expanse of wings 1 inch.

Upper side brown, much marked with bright fulvous,
which covers the central margin of the fore wings from
near the base to near the end of the cell and back to the
median vein, except a brown streak in the cell from the
base. A submarginal row of confluent spots extends
from the costa to the hind margin, broken opposite the
cell, two small spots ranging outside the line, with a
space between them and the costal spot. Edge of hind
margin also fulvous.

The hind wings have a similar row, or rather one
long spot or band, across the wing, and two spots on the
disk and on the costa. Fringes fulvous.

Under side pale brown washed with fulvous, which last
color prevails on the apical part of the fore wings and

on the anterior part of the hind wings. The spots of the upper side reappear, enlarged, and two spots near the apex of the fore wings are connected with the costal spots.

West Virginia, Colorado, California.

132. ANCYLOXYPHA NUMITOR, Fab.

Expanse of wings from .8 to 1 inch.

Upper surface of fore wings blackish brown, washed more or less with dull dark yellow; the cell dusky, but in front of the cell nearly clear yellow, and more yellow below the cell than in it. Hind wings dark yellow, costa and outer margin blackish brown.

Under side of fore wings brown, the costa and outer margin, to near the posterior angle, yellow. Hind wings uniform yellow.

Harris states that the chrysalis is rather long, nearly cylindrical, but tapering at the hinder extremity, and with an obtusely-rounded head. It is reddish-ash-colored, minutely sprinkled with brown dots.

Maine to Texas, Nebraska.

133. THYMELICUS POWESHIEK, Park.

Expanse of wings 1.2 inches.

Upper surface dark brown, the costal margin to near the apex dull yellow.

Under side of fore wings dark brown, the basal two thirds of costal edge dull yellow, the apex washed with yellowish white. Hind wings dark brown, the veins white; the whole surface, except between the submedian vein and near the inner margin, sprinkled with white scales. Fringes brown, basal half white all round.

Iowa, Illinois, Montana, Colorado.

134. Pamphila Massasoit, Scud.

Expanse of wings from 1.1 to 1.4 inches.

Male.—Upper surface blackish brown, with a slight violet reflection on the fore wings; three small, sub-obsolete, yellow, intervenular dots in a row near the costa beyond the cell, and occasionally a faint, small, yellowish spot or two on the middle of the hind wings. Fringe slightly paler, yellowish round the anal angle.

Under side of fore wings about the same color as the upper, the costal and outer margin tawny orange-yellow, the spots of the upper side repeated with more distinctness; a few scales in two small patches near the middle. Basal color of hind wings blackish brown, but so washed with tawny orange as to be almost that color, paler than the female. Across the wing, a little beyond the middle, is a series of six pale yellow spots; the first indistinct; the second nearly square, with the outer end rounded; the third oblong, reaching from near the margin to the middle of the cell, a faint brown bar at the end of the cell; the fourth and fifth oblong, one-third the length of the third; the sixth, like the first, subobsolete. Body, head, and antennæ, above, the color of the upper surface; below, pale whitish yellow.

Female.—Above, the markings of the under side show more plainly than in the males; the yellow of the under side is darker, and the fourth and fifth spots on the under side are pointed towards the base and somewhat blended with the third.

Food-plant and larva unknown.

Eastern and Middle States, Nebraska, Colorado, Texas.

135. PAMPHILA ZABULON, Bd.—Lec.

Expanse of wings 1.2 inches.

Male.—Upper surface pale dull yellow, the fore wings dusky at the base, the outer fourth blackish brown, dentate within; the same extending along the hind margin to near the base, and very narrowly edging the costa. There is an oblique bar at the end of the cell, and a patch beyond, almost touching the apical portion of the border; beyond this patch the border is narrowed, with a narrow line of yellow intervening. Hind wings bordered all around with blackish brown.

Under side of fore wings pale yellow, the basal half, except the costal margin, dark brown, the marks of the outer end repeated with less distinctness. The hind wings have a broad, pale yellow, slightly clouded band across the wing beyond the middle; the basal third, except a costal patch, reddish brown. The outer border contains several irregular pale brown spots; and there is a dark brown, yellow-washed streak below the submedian vein.

Var. HOBOMOK, Harr.—This form has the markings of the male much as in the typical form, but the dark spots are more pronounced, with often a subterminal patch above the discal bar, and the yellow is bright tawny. The hind wings have the yellow band of the under side narrower, leaving a continuous outer border, in width one-fourth the length of the wing, with some lilac scales.

The female of this form has the yellow a little paler than the ordinary male, the veins all brown, more dusky, suffusing over the basal half of the wings, with the yellow on the under side of the hind wings more contracted than in the male.

Dimorphic form female POCAHONTAS, Scud.—This form of female is similar to QUADRAQUINA, but duller in color, traces of a spot in the cell of the fore wings, the four posterior spots of the outer row not clearly defined. Hind wings a little pale in the middle. The marks on the under side are somewhat blurred, and the hind wings have an indistinct subterminal band.

Var. female QUADRAQUINA, Scud.—This has the upper surface the same shade as the outer border of the male, with a slight vinous reflection; beyond the middle of fore wings a broken row of pale yellow spots, three near the costa, then two nearer the outer margin, which are small and mostly oblong, then four to the hind margin, trapezoid, and all but the third larger. Hind wings without spots.

Under side blackish brown, the costal margin of both wings washed with tawny ochre, the apex of the fore wings whitish, the outer portion of the hind wings washed with lilac. The spots of the upper surface are repeated, the posterior four enlarged and more or less confluent.

Body dark brown, with greenish hairs above, paler beneath.

The eggs (Fig. 82) are pale green, nearly globular,

FIG. 82.

P. Zabulon,
egg, × 12.

somewhat flattened, and under a strong lens they appear reticulated over the surface with fine six-sided markings. These are deposited on grass, upon which the larvæ feed.

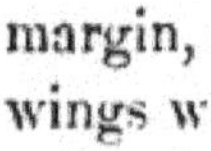

The young larva, which hatches from the egg in about ten days, is .1 of an inch long, with a large,

prominent, shining black head, and a creamy-white body,
with a yellowish tinge posteriorly.
The second segment is half circled
with a black line.

FIG. 83.

The larvæ station themselves on
the inside of the leaves, near the
joints, and, by drawing portions of
the leaves together with silk, form a rude case, in which
they secrete themselves.

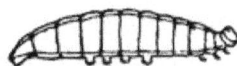

P. Zabulon, larva (natural
size).

Canada to the Gulf of Mexico, Mississippi Valley.

136. PAMPHILA SASSACUS, Harr.

Expanse of wings from 1.2 to 1.4 inches.

Male.—Upper surface dull dark yellow, the outer
border of fuscous or dark grayish brown, not quite one-
third the length of the wing, crenate within, the base
dusky. On the fore wings is a black, oblique stigma, or
sex-mark, below the cell, with a little fuscous below it,
and at its end a fuscous patch, which is separated from the
border by a few fulvous spots in a broken line; veins
fuscous. The yellow of the hind wings is surrounded
by a fuscous border.

Under side brownish fulvous, the posterior half of
fore wings fuscous; the five subterminal spots of the
upper surface repeated, a spot at the end of the cell, a
large triangular patch, with dentate outline, all yellowish
white. The hind wings have a subterminal band of
six whitish spots across the anterior two thirds of the
wing, and a spot at the end of the cell. One specimen
having a slight greenish-yellow wash to the under surface
has these spots indistinct.

Female.—This differs from the male in having the

u   26*

hind wings washed in the centre with yellow, the yellow area less than in the male and not clear yellow. Nearly all of the fore wing is dark grayish brown or fuscous, the basal half, or in some examples a little more, washed with yellow, or only the anterior edge and a portion of the base sprinkled with yellow scales. The fore wings have a row of eight yellow spots, of which 4 and 5 are out of line with the others, being nearer the margin; and two elongate spots, more or less distinct, in the outer part of the cell.

Under side grayish brown, a little pale, the hind wings and the anterior and outer portion of the fore wings well sprinkled with ochraceous scales; the spots of the fore wings repeated, pale, the lower of the outer row broadly expanded; the hind wings with a faint row of three or four spots beyond the cell.

Body brown above, with grayish hairs, lighter beneath.

The larva is said to feed on grass, the butterfly appearing on the wing in the middle of June.

New England, New York to Nebraska, Georgia, Florida, Colorado.

137. Pamphila Metea, Scud.

Expanse of wings from 1.2 to 1.35 inches.

Female.—Upper surface dark brown tinged ochraceous, especially on the hind wings. Fore wings with the following white markings: two small spots at the extremity of the cell; three small spots, one above the other, on the costal border, a little more than three-fourths the distance from the base; below these, and half-way between them and the outer margin, one above the other, two small spots; placed successively a little nearer the base than the last, two more spots, somewhat larger, between the branches of the median vein, and traces of a small one on the submedian.

Hind wings uniform in tint, with a faint ochraceous repetition of the markings beneath. The outer margin of both wings is narrowly edged with black, the fringes slightly paler than the upper surface.

Under side dark brown, on the hind wings approaching to black, with some grayish scales towards the outer border. The fore wings have the markings of the upper surface repeated with greater distinctness than above, and a large pale brown spot at the posterior angle.

Hind wings with a band formed of grayish-white spots between the venules, starting at the costa at two-thirds the distance from the base, nearly reaching the outer margin in the space between the subcostal and median veins, thence bent towards the inner margin at a little less than a right angle, terminating at the submedian.

Male.—This form scarcely differs from the female on the upper surface, except that the spots are a little more distinct and pale yellow, especially the row on the hind wings. Stigma, or sex-mark, oblique, narrow, black, broken, the upper part longer than the lower.

Under side as in the female. Body black above, with greenish hairs, below gray, with a few greenish hairs on the thorax; palpi yellowish white, gray at the tip.

Grass is the food-plant of this species.

It has been found in Connecticut, New York, Texas, and Colorado.

138. Pamphila Uncas, Edw.

Expanse of wings from 1.35 to 1.6 inches.

Female.—Upper surface fuscous, the base and posterior part of the fore wings, and a broad band through the hind wings from the base out, washed with dusky fulvous. The fore wings have a subterminal broken row of spots, all whitish but the last, which is yellowish, also a small spot at the end of the cell; the two between the branches of the median vein with the outer angles much extended. Hind wings with the subterminal spots of the under side showing through a little.

Under side fuscous gray, sprinkled with pale yellow scales. The spots on the upper surface of the fore wings are repeated, white in color, those of the subterminal row blended into three groups, the lower widened posteriorly, so as to suffuse most of the posterior angle area. The hind wings have two very much bent white bands, the outer not reaching the inner margin.

Male.—Upper surface fuscous, the fore wings, with the cell and a patch below the stigma, distinct yellow, inclining to fulvous, the hinder portion, from the end of the cell to the posterior angle, washed with yellow, and a row of five yellow spots in the outer fuscous field, the two beyond the cell much out of line with the others. The stigma very oblique, narrow, jet-black, contracted in the middle.

Hind wings, with all but a costal edge and a very narrow terminal border, heavily washed with yellow, inclining to a fulvous shade in the central portion. Fringes white, fuscous at base. Under side as in the female.

Larva and food-plant unknown, though the latter may be grass.

Delaware to Ohio, Dakota to Arizona.

139. PAMPHILA SEMINOLE, Scud.

Expanse of wings 1.35 inches.

Male.—Upper surface dark brown, slightly tinged with violet, the fore wings, with the basal half and costal edge, sprinkled with fulvous yellow; a broken row of dull yellow spots beyond the middle, consisting of three below the costa in line, two beyond the cell farther towards the margin, and three, larger than the others, between the branches of the median vein and above the submedian, each of these three reaching from vein to vein; a yellow bar at the end of the cell. Stigma black, narrow, broken near the middle, the parts slightly curved, the ends overlapping each other a little.

Hind wings with brownish-yellow hairs, and a row towards the outer margin of about five small yellow spots.

Under side scarcely paler than the upper, the hind wings more tinged with reddish, sprinkled with yellow scales which are pale on the hind wings, more distinct fulvous yellow along the costal edge of the fore wings and near the anal angle of the hind wings. Spots of upper side repeated, but paler, almost white, spot 7 of the fore wings enlarged, and spot 8 shading out on each side.

Female.—Similar to the male, but scarcely sprinkled with yellow; lacking the stigma; the spots a little more prominent. Under side as in the male.

The larva and food-plant are unknown.

It has been found in Florida, North Carolina, New Jersey, and Iowa.

140. Pamphila Leonardus, Harr.

Expanse of wings from 1.3 to 1.4 inches.

Male.—Upper surface blackish brown, more or less overlaid with fulvous yellow scales from the base to the outer third of the fore wings, with a broken subterminal row of clearer yellow spots; three of these below the costa in line, two beyond the cell farther out, and separated from the cell by a space without yellow, and two or three below, in line with the first, continued by an oblique shade to the hind margin; also a distinct spot at the end of the cell. Stigma black, oblique, concave below. The hind wings have a curved band not reaching either margin; an obscure spot in the cell, and greenish-yellow hairs over the inner half of the wings.

Under side bright reddish brown, the fore wings blackish from the cell to the hind margin, but not reaching the outer margin except at the posterior angle. The spots of the upper surface are repeated, but the spots below the cell are blended and enlarged into a subtriangular patch. The hind wings have a small spot at the end of the cell, and beyond a curved row of six or seven spots.

Female.—This form has the basal third of the fore wings only moderately sprinkled with yellow scales, and the stigma is absent. The spots in the outer row are larger than in the males. Under side like the male.

According to Mr. Scudder, this species feeds on grass in the larval state.

New England to West Virginia; Indian River, Florida; Kansas.

141. PAMPHILA MESKEI, Edw.

Expanse of wings from 1.5 to 1.6 inches.

Male.—Upper side dark brown, marked and spotted with reddish fulvous; three small spots in subcostal interspaces of fore wings, two others opposite the cell and towards the outer margin, and in line with these an oblique row crossing the median interspaces, the lower spots merged in the fulvous of disk or cell to base and anteriorly to costa; the stigma in two sections, the anterior one completely crossing the lower median interspace, a narrow, smooth, black, slightly bent ridge; the lower one in submedian interspace parallel to the line of the other, short, not reaching either venule; behind the stigma is a narrow, blackish, rough patch. The hind wings have the central part obscure fulvous, with an imperfectly defined series of spots between the cell and the outer margin. Fringes of the fore wings fuscous next the margin, whitish outside; of the hind wings, fuscous next the margin, then fulvous, and outside whitish.

Under side of hind wings bright ferruginous, of one shade, and without spots; apex of fore wings bright ferruginous, the outer margin a little obscured, the spots indistinctly repeated, next the base and against the stigma black.

Female.—Color dark brown; the fore wings have spots placed as in the male, distinctly defined throughout, the basal area being brown, instead of fulvous as in the male. Hind wings as in the male, though the spots may be more clearly defined.

Under side as in the male, but on the disk of the hind

wings is an indistinct bent row of small paler spots corresponding to the spots of the upper side.

Body above black, covered with fulvous hairs; thorax below yellowish; abdomen the same, with a fulvous tint; palpi light yellow, with a fulvous tint; antennæ black above, yellowish below; club black on both upper and under surface, on the sides fulvous.

Larva and food-plant unknown.

Texas; Indian River, Florida.

142. PAMPHILA HURON, Edw.

Expanse of wings from 1.2 to 1.5 inches.

Male.—Upper surface of fore wings dark yellow tinged with fulvous, dusky at base, the veins fuscous, and a fuscous outer border about one-fourth the length of the wing, crenate within, narrowed opposite the cell, before which there is a fuscous patch. Stigma somewhat quadrate, the upper outer angle produced, velvety black above and below, with a nearly round blackish patch beyond.

Hind wings with the central portion of the wing yellow, but washed a little with dusky, a continuous fuscous border round the wing.

Under side dull yellowish fulvous, the hind wings and terminal portion of the fore wings tinged with grayish; the basal half of the fore wings from the cell to the hind margin, and a border to the posterior angle, blackish. The fore wings have a faint subterminal band, much as on the under side of *P. Leonardus*. Base of hind wings dusky, a faint subterminal band.

Female.—This has the upper surface blackish fuscous, the fore wings washed with yellowish from the base to

the outer third along the costal and hind margins; a
black space in place of the stigma of the male; the usual
broken subterminal row of spots, the first three and the
last two translucent. Hind wings with yellowish scales,
and greenish hairs over the inner half, an indistinct spot
in the cell, and a band beyond not reaching either margin,
in width about one-fourth the length of the wing.

Under side like that of the male in color; the row
of spots of the fore wings repeated, the hind wings with
two much curved rows of white confluent spots.

Atlantic States to Florida, Mississippi Valley, Texas,
Arizona.

143. PAMPHILA PHYLÆUS, Drury.

Expanse of wings from 1.1 to 1.3 inches.

Male.—Upper surface yellow, dusky at base along the
veins, and on the costal and inner margin of the hind
wings. Fore wings with an outer border of eight cuneiform fuscous spots, the length being about one-fourth
the length of the wing, the third and fourth from the
apex about half as long as the others, and the eighth
half-wedge-shaped. Stigma slender, oblique, velvety
black, with a fuscous patch below; a bar at the end
of the cell, with two rays from it. Hind wings with a
border of five cuneiform spots, the second very short.

Under side paler than the upper, and the yellow not
so bright. Fore wings fuscous below the cell and along
the hind margin, the stigma and part of the terminal
spots showing, but not those at the apex. The hind
wings have two subobsolete rows of fuscous spots, the
inner across the middle of the wing.

Female.—Upper side dark brown, with a little vinous

o   27

reflection. The fore wings have the basal third washed
with dull yellow, the spots brighter dark yellow. There
is a spot at the end of the cell, and an oblique broken row
behind; the first three in the subcostal interspaces are oblong, slightly narrowed at the base; the next two opposite the cell, farther towards the margin, small and round
or quadrate; number 6 like 2 and 3; number 7 quadrate; number 8 a mere point; number 9 triangular, and
extending as a shade along the hind margin to the base.
Hind wings overlaid on basal and inner portions with
dull yellow hairs, and with a subterminal row of spots
much as in the female of *P. Huron*.

Under side darker yellow than in the male, the fuscous covering more than the posterior half of the wing;
the spots of the upper side repeated, whitish.

According to Dr. A. W. Chapman, the larva of this
species when full grown is .7 of an inch long, fusiform,
of a uniform dull green color, and thickly granulated
with pale points. The collar on joint 2 is dark brown.
Head small, dark brown.

The chrysalis is .5 of an inch long, nearly cylindrical,
pubescent. Color pale green; a black line, interrupted
on the posterior joints, extends from back of the head-
case to the last joint, with a lateral black streak on the
thorax, and a row of spots on the abdomen. More or
less punctured throughout.

Food-plant, grass.

Middle and Gulf States to the Pacific.

144. PAMPHILA BRETTUS, Bd.—Lec.

Expanse of wings 1.1 inches.

Male.—Upper surface yellow, with a fuscous terminal

border crenate within, covering about one-fourth the length of the wing, receding somewhat opposite the cell of the fore wings; the base dusky; the veins of the outer half of the fore wings fuscous. The stigma has the central streak dull black, with velvety black each side of this, and a small blackish patch below. A fuscous streak extends from the lower end of this to the base along the submedian vein, and another irregular-shaped patch extends from the upper end of the stigma so as almost to touch the terminal border opposite the cell.

Under side dark brown, overlaid with yellow; costal portion of fore wings yellow, basal portion fuscous. An irregular subterminal yellow band crosses both wings.

The egg is white, smooth, hemispherical. It is deposited on grass (*Paspalum setaceum*).

The young larva is white, with a large black head, and black collar. When full grown it is one inch long, pale green, with a dark dorsal stripe, and an obscure line on each side. Collar black, on each side a black dot separated from the collar. Stigmata black. Head rounded, projecting obliquely, granulated with black, the sides of face and two streaks on upper part of face yellow-white.

Chrysalis .75 of an inch long. Color pale green, the abdomen whitish; wing-cases smooth, finely veined; the antennæ-case extending in a filiform point to the end of the abdomen. On each side of head-case a dark point, and a row of dark points along the sides of the abdomen.

Gulf States, West Virginia.

145. PAMPHILA OTHO, Sm.—Abb.

Expanse of wings from 1.2 to 1.25 inches.

Upper surface dark brown, with a little vinous re-

flection. The fore wings have a series of yellow spots,
—the first three-anteapical, the next two between the
branches of the median vein, and the sixth a small one
on the submedian. The hind wings have the usual
greenish-yellow hairs over the inner part, and two small,
faint, contiguous spots at the end of the cell. The male
differs from the female in having an oblique stigma, and
in having an overlaying of greenish-yellow scales on the
basal portion of the wings, though scarcely enough to
change the color.

Under side yellowish brown, the posterior half, or
more, of the fore wings blackish. The spots of the fore
wings are repeated, except the last one. The hind wings
have an indistinct yellowish band of five or six spots.

Var. EGEREMET, Scud.—This is similar to the other
form, but differs in wanting the marks on the upper side
of the hind wings, and the anterior and posterior spots
of the fore wings are lacking.

The under side is dark blackish brown, obscure at the
base, the hind wings, especially of the male, sprinkled
with olivaceous scales, the posterior part of the fore
wings more blackish, and the costal margin sprinkled
with yellow. The spots of the fore wings are repeated,
and there is an indistinct row of spots across the hind
wings. There is the same difference between the males
and the females on the upper surface as in the form *Otho*.

This is a variable species, the form *Otho* seeming to
be the Southern form, while *Egeremet* is the Northern,
with intergrades covering the intervening ground. One
of these forms, a female, named *Ursa*, Worth., differs
from the form *Otho* in having the posterior spot on the
fore wings lacking, and in there being traces of a band

of elongate spots on the upper side of the hind wings, seen fairly only in certain lights.

Body above dark brown, lighter beneath.

Atlantic States, Mississippi Valley, Texas.

146. PAMPHILA PECKIUS, Kirby.

Expanse of wings 1 inch.

Female.—Upper surface dark blackish brown, the markings dark yellow, slightly fulvous-tinted; there is a slight sprinkling of yellow scales over the surface, especially the basal portion. The markings of the fore wings are: a slight ray in the upper part of the cell, and traces of one in the lower part; and a broken band of seven spots beyond the middle, the two opposite the cell beyond the line of the others. The first three of these spots are oblong in the subcostal interspaces; the next four mostly quadrate; sometimes the fourth of the seven is wanting, and there are occasionally a few scales on the submedian vein in line with the last spots. Hind wings with a band of five spots, in width about one-fourth the length of the wing; spots 3 and 4 the largest.

Fig. 84.

Pamphila Peckius, under side.

Under side fulvous brown, the basal half of the fore wings fuscous; the spots of the fore wings are repeated, lighter yellow, there being eight of them instead of seven. The hind wings have two broad pale yellow bands, a little irregular in outline, the lower part of the inner often united with the middle of the outer, as in Fig. 84.

Male.—Of the same color as the female, the basal two thirds of the fore wings heavily washed with yellow;

the outer two thirds of the cell, and the space in front
of that, nearly clear yellow; below the cell an oblique,
sinuous, velvety-black sex-mark; and below this a nearly
round brownish-olive patch. The outer third is sprinkled
with yellow scales, and contains the outer row of spots
found on the wing of the female, the seventh spot partly
lost in the olive patch. The hind wings are similar to
those of the female, but are sprinkled with yellow.
Under side the same as in the female.

According to Professor Fernald, this species feeds on
grass. The eggs are pale greenish yellow, strongly convex above, and flattened at the base, and the surface is
faintly reticulated. They hatch in fourteen days. The
young larva is .1 of an inch long, with a large shining
black head. The body is dull brownish yellow, dotted
with black, with a ring of brownish black on the second
segment. Under side paler than the upper, and the whole
surface clothed with fine hairs. The butterfly is on the
wing from June to July.

New England to Wisconsin, Illinois, West Virginia,
Kansas.

147. PAMPHILA MYSTIC, Scud.

Expanse of wings from 1.1 to 1.2 inches.

Male. — Upper surface yellow, slightly brownish-
tinted; an outer border of dark blackish brown, about
one-fourth the length of the wing, not crenate on its inner
edge, but receding a little opposite the cell of the fore
wings and at the apex; base dusky. Stigma oblique,
black, slender, with a blackish patch below it, and an
irregular patch from the end of the cell outward, the
corners of which connect with the outer border, leaving

a small yellow spot enclosed. This patch, the stigma, and the dusky base form a continuous line.

The hind wings have the outer border narrower than the border of the fore wings, but have broad inner and costal borders; veins dark, with some shading at the end of the cell.

Under side somewhat paler than the upper, the fore wings blackish below the cell and along the hind margin, a band of paler yellow beyond the middle, the posterior spots expanded. Hind wings with a broad subregular subterminal band, and a patch in the cell; all indistinct.

Female.—The fore wings dark brown, the outer two thirds of cell pale yellow, and some yellow suffusion in front of this. There is the usual row of spots marking the outer third: the first three oblong; the fourth obscure; the fifth triangular; the sixth oblong; the seventh subquadrate, convex within, concave without; the eighth irregular. The hind wings have a patch at the end of the cell, and a band of five spots beyond, the first a little out of line with the others; the ground color the same as that of the fore wings.

The under side is marked as in the male, but the surface is more fuscous, except the anal portion of the hind wings and the anterior basal portion of the fore wings, which are but little darker than in the males.

Like many other species of Pamphila, this feeds on grass in the larval state. The eggs are, according to Mr. Scudder, of a pale yellowish-green color, strongly convex above, and with the base flattened. The surface appears smooth under a lens, but under a power of eighty diameters is seen to be faintly reticulated. The egg period lasts eight or ten days. The young larva is .1 of an inch long, with a large shining black head, and a white body tinged with yellowish brown, this tinge being more apparent towards the posterior part.

The full-grown larva is of an oval outline; the head not large in proportion to the size of the body, but prominent and much larger than the second segment; it is of a dull reddish-brown color, edged with black on the hinder part, and clothed with minute whitish hairs. The body is dull brownish green, with hairs similar to those on the head; a dorsal line and numerous dots over the surface of the body are of a darker shade. Joint 2 is pale whitish, with a line of brownish black across the top. The last joints are paler than the rest, and the under side of the body is paler than the upper.

So far as known, there is only one brood in a season, and the butterflies are on the wing in June and July.

New England to New York.

148. Pamphila Cernes, Bd.—Lec.

Expanse of wings from 1 inch to 1.1 inches.

Male.—Upper surface dark olivaceous brown, with a little vinous reflection; fore wings, with the cell, the costal area to half-way between the cell and the apex, and a patch beyond the upper end of the stigma, clear yellow. The whole area below the cell, except a dusky patch outside the stigma, washed with yellow. Stigma oblique, velvety black, contracted a little in the middle. Hind wings with olive hairs and sprinkled with yellow scales.

Under side blackish or fuscous, the posterior part of the fore wings clear, the outer half of the anterior portion of the fore wings and all of the hind wings overlaid with yellow; the cell and costal margin before the cell of the fore wings clear yellow. The fore wings have five spots marking the outer third, the three costal obscure, the two between the median venules pale yellow, the lower excavate externally.

Female.—Upper surface of the same ground color as in the male, the fore wings with a ray of clear yellow in the cell, and the basal half sprinkled with yellow scales, more so on the costal and hind margins. The five yellow spots that are on the under side of the fore wings of the male are distinct on the upper surface of the female, with some scales on the submedian vein in line with the others. Hind wings as in the male, but not sprinkled so heavily.

Under side as in the male, but not so heavily overlaid with yellow.

Body dark brown above, with greenish hairs; a little lighter beneath.

The larva is unknown.

New England to Montana, Florida.

149. PAMPHILA MYUS, French.*

Expanse of wings .95 of an inch.

Male.—Upper surface dark olivaceous brown, with a slight vinous reflection, about the same shade as *P. Cernes*, which it much resembles. The fore wings have the discal cell and the area in front of the cell like *Cernes*, heavily washed with yellow of a little darker shade than in that species, the same color extending beyond the cell along the costal area three-fourths the distance from the base to the outer margin; below the cell the same shade

---

* Mr. E. M. Aaron thinks this is *P. Baracoa*, Luc.

v

of yellow extends along the median vein the same distance, the area below this to the hind margin rather heavily sprinkled with yellow scales, except the space beyond the lower half of the stigma, being in this much like *Cernes*.   In *Cernes* there is a quadrate sinus of the terminal dark brown of the wing dipping into the yellow beyond the cell, coming up to the cross-vein.   In this species the sinus is of the same width, but extends inward above the median vein, ending in a point halfway to the base of the wing.   The stigma is black, narrow, oblique, entire, though constricted below the middle, shorter than in *Cernes*, does not reach the submedian below, and the upper end reaches only the second branch of the median, while in *Cernes* it passes beyond this venule, the lower third bent a little towards the base, not more than half as wide as in *Cernes;* below the stigma an oblong patch of blackish scales which are bronzy in certain lights.   The hind wings are sprinkled with yellow scales, the inner half with yellowish hairs which are less olivaceous than in *Cernes*.

One specimen has on the fore wings, marking what is above described as the outer boundary of yellow, five small yellow spots which are paler than the yellow along the costa,—three in a line back from the costa, and two in the median interspaces; the yellow washing does not quite reach to these spots, there being less yellow also at the base; varying in amount of yellow, as is sometimes seen in different specimens of *Cernes*.

Under side of fore wings much as above, the yellow orange-tinted, the row of slightly paler spots at the end of the yellow showing more distinctly than above, the apical half of the terminal space sprinkled with yellow,

the posterior half of the wing blackish, the sinus beyond
the cell heavily sprinkled over.

Hind wings dark brown, with a vinous reflection,
sprinkled with pale yellow scales, a narrow discal band
of small confluent whitish spots marking the outer third,
much as in the species of Amblyscirtes, not very distinct.

Female.—This lacks the stigma of the male, is marked
above much as the female of *Cernes*, but is of a darker and
brighter yellow, the whole area in front of the cell and
to the anteapical spots nearly clear yellow, the rest of
the basal two thirds sprinkled with yellow, much as in
the male. On the under side the obscure band on the
hind wings is a little more distinct than in the male.

Body concolorous with the wings above, the thorax
with olivaceous hairs, the abdomen sprinkled with yellow; beneath yellowish white, about the shade of
Cernes.

Florida, specimens obtained during the summers of
1883 and 1884.

150. PAMPHILA MANATAAQUA, Scud.

Expanse of wings from 1 inch to 1.3 inches.

Male.—Upper surface dark brown, with a brassy reflection. The fore wings have, about three-fourths the
distance from the base, two or three small yellow intervenular spots in a line back from the costa, and a series
of three more below these,—two between the branches of
the median vein and one above the submedian; the first
of these nearly square, the second oblong, twice as broad
as long, the third small. The stigma is black, narrow,
oblique, extending from the forking of the median at

the end of the cell to the submedian. Hind wings without marks, the hairs on the inner half brown and olivaceous.

Under side tawny yellowish brown, the fore wings with the spots as above, but paler, the one above the submedian shaded out considerably with white. Hind wings with a transverse row of four subobsolete pale yellow spots.

Female.—The same color as the male, the spots in the discal row of fore wings larger and more distinct, the one above the submedian somewhat hour-glass-shaped; base of fore wings sprinkled with yellow, the cell washed with the same. The stigma absent.

Under side as in the male, but the surface more sprinkled with pale yellow, the washing of yellow on the fore wings repeated, the spots the same as in the male, but a little more distinct.

Body dark brown above, with olivaceous hairs, below yellowish white.

United States generally.

151. PAMPHILA VERNA, Edw.

Expanse of wings 1.25 inches.

Male.—Upper surface dark brown, with a slight purplish reflection. The fore wings have the costal edge slightly sprinkled with yellow; the small, yellowish, translucent, intervenular spots in a line back from the costa, nearly three-fourths the distance from the base to the apex, and two larger spots between the branches of the median vein, the second twice as large as the first; a few scales are seen above the submedian in line with these, and there is a small spot in the lower part of the cell near the end. Stigma black, oblique, narrow, some-what constricted, but not broken. Hind wings without spots, hairs yellowish green. Fringes yellowish gray.

Under side about the same color as above, all but the posterior part of the fore wings and a stripe within the inner margin of the hind wings tinged with yellow bronze having a purplish reflection. Spots on the fore wings repeated; the spot on the submedian considerably enlarged.

Hind wings with a faint discal row of five small whitish spots.

Female.—This differs little from the male in mark-ings and color; there is less sprinkling of yellow scales, and the few scales above the submedian may be absent. Under side as in the male. Stigma absent from the fore wings above.

Body concolorous with the wings above, gray be-neath.

The larva and food-plant are unknown.

New York, Maryland to Georgia, West Virginia, Ohio, Indiana, Kansas.

152. PAMPHILA VESTRIS, Bd.

Expanse of wings 1.28 inches.

Male.—Upper surface dark glossy brown, as in *P. Metacomet;* outer margin blackish brown; fringes dark brown. Fore wings with some dull yellowish scales on the inner half of the costa, on the outer side of the stigma. and within it, between the median and submedian veins, Stigma velvety black, consisting of two acutely ellipsoidal spots, which join on the lower median venule and have their extremities resting on the submedian and second

28

branch of the median ; the inner spot with distinct black scales near the submedian vein.

Under side brown, blackish over the stigma, with obscure yellow shades exterior to it as the only markings. Abdomen above concolorous with the wings, with yellowish scales laterally. Thorax beneath, and abdomen contiguous, brown, with some longer clay-colored hairs. Palpi clothed with bristling yellow scales, from which the tip of the last joint barely projects.

Female.—Fore wings with dull yellow scales and hairs, more numerous on the inner half of the hind margin, and nearly absent from the outer margin ; two yellow spots between the branches of the median vein, the outer one scarcely more than a dot, the inner sub-quadrangular ; no anteapical spots, but in their place some clustering yellow scales.

Under side dark brown, the fore wings reddish brown basally, and the hind wings of the same shade throughout, except towards their inner margin. The two spots of the upper surface of the fore wings are reproduced beneath somewhat more obscurely. Thorax and front of head clothed with yellowish scales ; palpi with black scales above, and beneath with some clay-colored scales.

California, Colorado, Indian River, Florida.

153. PAMPHILA METACOMET, Harr.

Expanse of wings from 1.1 to 1.3 inches.

Male.—Upper surface dark brown, slightly glossed with greenish yellow above; the usual oblique velvety-black stigma. The under side of the wings slightly paler, the hind wings with a transverse row of four very faint yellowish dots, which, however, are often wanting.

Female.—Of the same color as the male, lacking the stigma, and having two yellowish dots between the branches of the median vein, and two more anteapical near the costa beyond the cell. The under side has the spots of the upper surface reproduced; hind wings as in the male.

This butterfly is to be found in July, the larva feeding on grass.

New England to Montana, Kansas, Nevada.

154. PAMPHILA ACCIUS, Sm.—Abb.

Expanse of wings 1.4 inches.

Upper surface dark blackish brown, slightly olive-tinted. Males with an oblique black stigma, with a white dot at the upper end of it in the upper median interspace, and three small white dots in the subcostal interspaces beyond the cell. The females lack the stigma, but have besides the anteapical spots an oblique row of three in the median interspaces and above the submedian, the middle spot the largest.

Under side dark reddish brown, the posterior part of fore wings and inner part of hind wings blackish. The following parts are suffused with lilac: the outer part of the fore wings, narrowing from the middle to both margins; a similar space on the hind wings, also a patch in the middle. The white spots of the upper surface are repeated.

Some specimens have a white spot at the end of the cell of the fore wings besides those mentioned.

The mature larva is 1.33 inches long, slender, nearly white, but under the lens mottled and dotted with darker lines and points, the rings on the posterior half of each joint more prominent and less dotted; collar black. Head rather small, oblique, oval, flattened frontally; white, with a black band around the top and sides, a black streak down the middle of the face, and a short black streak on each side of this last, not reaching the band at top.

The chrysalis is smooth, white, the head-case tapering into a slender pointed beak.

The larva was found in August by Dr. A. W. Chapman wrapped in the leaves of a large grass (*Erianthus alopecuroides*).

Gulf States, North Carolina, Eastern Pennsylvania, Southern Illinois.

155. PAMPHILA LOAMMI, Whitn.

Expanse of wings from 1.3 to 1.5 inches.

Male.—Wings above dark glossy brown, darker basally. Fringes light brown, with a blackish line at extreme base. Fore wings with a subcostal transverse row of quadrate whitish spots, situate one in each of the three terminal subcostal interspaces near the base; the upper one one-half its length nearer apex. A large subquadrate spot crossing second median interspace at one-third the distance from its base. An obsolescent transverse line in lower median interspace, equidistant between its base and spot in second interspace. A narrow black stigma broken by the lower branch of the median vein; upper portion straight, commencing at second branching of median and crossing the interspace to first median venule near its source. The lower portion of the bar commences below the venule about its own width removed outwardly, is strongly concave within, and reaches the submedian vein about two-fifths its distance from the base. Hind wings without spots.

Under side dark chestnut-brown. Apex of fore wings and border of hind wings with a bloom of pearly scales. Fore wings with the markings of the upper side repeated, and two minute dots in subcosto-median interspaces, resting one on each venule; one in first median interspace and a transverse line in third. These five, including one in second interspace repeated above, are in line from apex to internal margin.

Hind wings with a curved sub-basal row of three small irregular white spots. The first is in the costo-subcostal interspace one-fourth the distance from its base, the second in the cell, and the third on the sub-median vein. A subterminal sinuate row; the first double, situate in the costo-subcostal interspace midway between its other spot and its extremity. A black streak extends from this spot sharply outward to the next spot below the subcostal vein, which is followed in the succeeding interspaces by five small transverse spots. All the spots of the hind wings have a black border.

Female.—General coloration a little lighter than in the male. Fore wings above with two spots at extremity of cell. An irregular transverse band commencing with three subcostal spots, the upper one not removed outwardly as in the male; the fourth twice its own width nearer the margin; the fifth in line with the first three; the sixth twice its width nearer the base; the seventh largest, removed its width internally; the eighth double or with upper half obsolete.

Under side of fore wings with upper markings repeated. Hind wings with basal row inconstant. First three spots of subterminal row running towards the outer margin; the others running at a right angle from the third, towards the inner margin. In one female example the subterminal row of hind wings is indicated above by a few lighter scales.

The larva and food-plant are unknown.

Taken in Florida and North Carolina.

156. PAMPHILA MACULATA, Edw.

Expanse of wings from 1.4 to 1.5 inches.

Female.—Upper surface uniform dark brown. The fore wings have three small round spots in the subcostal interspaces beyond the cell, extending from the costa back; two more, of larger size, in the median interspaces; and a third below these on the submedian vein, the latter sometimes obsolete. The hind wings have a small spot on the middle, in some examples obsolete. All these spots are semi-transparent, yellowish.

Under side nearly as above, fore wings washed with white near the posterior angle, spots of fore wings as above. The hind wings have three spots in a transverse row across the disk. Body black; palpi yellowish.

The larva, when full grown, is one inch long, slender, pale green, finely pubescent, the last two joints deeper green, collar light brown. Head oval, oblique, densely pubescent, slightly granulated, light brown.

The chrysalis is .8 of an inch long, cylindrical, dull green; pubescent, especially about the head. Head-case blunt, wing-cases smooth. On joints 8, 9, and 10 are two flat tubercles on the ventral side. Anal hook broad, triangular.

Gulf States; occasional in New York.

157. Pamphila Panoquin, Scud.

Expanse of wings from 1.2 to 1.3 inches.

Male.—Upper surface brown, with a bronze lustre. The fore wings have two of the usual three anteapical spots, those present representing the second and third of the usual number; subquadrate, the outer corners with a tendency to extend outward in points. There are usually one or two beyond the cell nearer the margin than the anteapical spots; when both are present the lower one is much the larger and a little nearer the cell. Below these is an oblique row of three spots,—two in the median interspaces and one in line on the submedian vein, the second a little more than twice as large as the upper, the one on the submedian often small; also a small spot in the lower side of the cell at the outer end. All these spots are pale yellowish. Stigma small, oval, parallel to the costa.

Under side of nearly the same color as the upper, much powdered with bronze scales on the costal margin of the fore wings, and at the base and along the veins of the hind wings. The spots of the fore wings show more distinctly pale yellow. On the outer part of the hind wings there is a white stripe following one of the discal venules, with a spot below and occasionally one above.

Female.—Without the stigma, colored and marked like the male, but usually lacking the upper spot beyond the cell.

The larva and food-plant are unknown.

Gulf States; Atlantic City, N.J.

158. PAMPHILA OCOLA, Edw.

Expanse of wings 1.4 inches.

Male.—Upper surface dark brown, slightly bronzy; the fore wings with one or two small spots in the sub-costal interspaces, the first one of the usual three being absent, often the second also, those present being mere points. There are three other spots in a bent row,—two in the median interspaces and one on the submedian vein, the one on the submedian so far towards the margin as to be out of line with the other two; the first or upper about one-fourth as large as the second, the second concave on the outside. All these spots are dull, dusky, translucent yellowish.

Under side of about the same shade as the upper, the spots of the fore wings repeated. The costal margin and most of the outer margin of the fore wings, and the veins of the hind wings, somewhat bronzy.

Female.—Color and markings the same as in the male, the second of the three oblique spots more than three times as large as the first. The bronze on the costal margin of the under side of the fore wings is very distinct.

Larva and food-plant not known.

Gulf States; Eastern Pennsylvania; Whitings, Ind.

159. PAMPHILA ETHLIUS, Cram.

Expanse of wings from 2 to 2.15 inches.

Male.—Upper surface dark blackish brown, some yellow scales over the basal part of both wings. The fore wings with seven whitish, translucent spots, as shown in Fig. 85. There are two in the two lower subcostal in-

ter-paces, the upper of the usual series absent; one in
the second interspace below these, beyond the cell, in
line with the first two; the fourth and fifth in the
median interspaces; the sixth on the submedian vein;
the seventh on the lower side of the cell near the outer
end. The first of these is oblong, the second quadrate,
the third oblong (the long way transverse to the wing
instead of longitudinal), the first and third nearly twice
as large as the second. The fourth is a little less than

FIG. 85.

Pamphila Ethlius (natural size).

half as large as the fifth, both somewhat trapezoidal,
concave on the outer side; the sixth is about the size of
the fourth, concave on the inside, the outside rounded;
the seventh is oblong, rounded outwardly and concave
inwardly. The hind wings have three marks similar to
the fourth on the fore wings, though not quite so large,
the first or anterior one often double. The fringes are
fuscous, darker at the base. The outer margin of the
hind wings is slightly excavate near the middle, below
which it is a little produced, somewhat like *Eudamus
Tityrus*, but broader and not so prominent.

Under side ochraceous brown over the hind wings,
and on the fore wings the costal margin and apical and

outer portion, ending in a point at the posterior angle, the rest of the fore wing blackish. The spots of the upper surface are repeated.

Female.—Like the male, except that the first three spots on the fore wings are more nearly of the same size.

Body concolorous with the wings above, with yellowish hairs, gray beneath.

Larva and food-plant unknown.

Gulf States; occasional in New York.

160. Pamphila Bimacula, Gr.—Rob.

Expanse of wings from 1.2 to 1.5 inches.

Male.—Upper surface dark brown, with a slight purplish tinge; the basal half of the fore wings washed with yellow inclining to fulvous, more apparent along the basal third of the costa and on both sides of the stigma, where it is nearly clear yellow. Stigma oblique, velvety black, broken in the middle into two narrow elliptical parts. There is a small pale spot in the first median interspace, and a few pale scales beyond the upper part of the cell in the place of the usual second and third anteapical subcostal spots. Hind wings without spots, but the middle and basal areas with olivaceous yellow hairs.

Under side grayish brown; the basal half of the anterior part of the fore wings, and the anterior part of the hind wings, heavily washed with yellow tinged with ferruginous; the rest of the hind wings, except the inner margin, and the outer part of the fore wings, well sprinkled with the same. The posterior part of the outer margin of the fore wings lacks these scales, and the basal half of the hind part is black. The fore wings have two pale spots in the median interspaces, the upper whitish and about one-third the size of the lower. Inner margin of hind wings sprinkled with gray. Fringes white, gray at the base.

Female.—Upper surface of the same general color as in the male, with very few of the yellow scales, and those mostly along the costa of the fore wings. There are two pale spots in the median interspaces; the anteapical scales as in the males. The hind wings have fewer of the olive-yellow hairs. Under side as in the males, with less gray on the inner margin of the hind wings.

Body black, the hairs of the thorax of nearly the same color as the yellow on the fore wings, those on the abdomen olive-yellow. Under side of body and palpi white.

Larva and food-plant unknown.

The butterfly is found in July from New England to Nebraska; Illinois.

161. Pamphila Pontiac, Edw.

Expanse of wings from 1.25 to 1.4 inches.

Male.—Upper surface dark blackish brown or fuscous, the basal two thirds of the fore wings so heavily washed with rather dark yellow as to make it clear yellow, separated by the brown veins in the cell, beyond the stigma and in the subcostal interspaces; the base of the wing and the bases of the subcostal interspaces having but little of the yellow, as also the area below the submedian vein. Stigma oblique, rather broad, velvety black, broken by the lower median venule into two elliptical parts which join by their oblique ends; the upper end stopping at the second branching of the median, the lower on the submedian about one-third the distance from the base.

Hind wings with the central area yellow, consisting of a broad band across the disk composed of four oblong spots between the veins, the second wholly or partially divided into two spots, and a small spot in the end of the cell; the width of the band about one-third the length of the wing. The inner third with yellowish olivaceous hairs.

- Under side dark brown, slightly ferruginous, the fore wings, with the basal two thirds, fuscous, shading outwardly into the brown. The costal margin is overlaid with yellow which extends into the cell. Marking the outer third is a band of dull yellow spots,—the two anterior in the two lower subcostal interspaces, the third in the second space below these, the next two in the median interspaces, and the sixth below the fifth, separated only by the vein; the last four forming a continuous band but for the brown veins. The hind wings, as well as the apical portion of the fore wings, are sprinkled with ferruginous yellow, the band of the upper surface repeated, but the spots somewhat contracted.

Female.—Upper surface dark or fuscous brown, as in the male, with slight vinous reflection, the base a little sprinkled with yellowish olivaceous scales. Two-thirds the distance from the base is a band of eight more or less distinct yellow spots, the sixth and seventh a little pale. The first three of these spots are in the subcostal interspaces, twice as long as broad; the next two beyond the cell, subquadrate, the fifth with its outer side in line with

the first three, the fourth with its inner side in the same
line; the sixth and seventh in the median interspaces,
much larger than the others, the outer sides excavate;
the eighth less distinct, and somewhat hour-glass-shaped.
The hind wings have a band through the middle, as in
the males, but the spots are smaller and the spot in the
cell is absent.

Under side as in the male. Body concolorous with
the wings above, under side pale yellow.

Larva and food-plant unknown.

Massachusetts to Nebraska; New Jersey.

162. PAMPHILA DION, Edw.

Expanse of wings from 1.4 to 1.6 inches.

Male.—Upper surface almost a copy of *P. Pontiac*,
but differs in the space above the cell of the fore wings
being more dusky, less washed with yellow, the base a
little more dusky, the cross-bar at the end of the cell
more distinct, the space beyond the cell as far as the yellow extends being nearly filled with clear yellow, while
in *Pontiac* the upper half is dusky and the stigma is
narrower, and there is more yellow below the submedian
vein. The stigma is oblique, velvety black, divided in
the middle into two elliptical portions, the upper slightly
the longer, of medium width, the two parts not touching
each other, there being a more distinct separation than in
Pontiac. The area on both sides of the stigma is clear
yellow, of about the same shade as in *Pontiac*. The
hind wings have the yellow band or area broader than
in *Pontiac*, the first spot less prominent, the second
longer, reaching well into the cell, without any dividing
cross or longitudinal marks, the remaining three spots

not very distinct, more obscured by the brownish-yellow hairs than are those in *Pontiac.*

Under side ochraceous yellow, tinged with brown, especially the costal and apical portions of the fore wings and the greater part of the hind wings; the posterior part of the fore wings blackish. The spots are obscure in the brownish-ochre ground, but are distinct in the blackish portion, pale yellow. They are two subcostal, a few scales, beyond the cell, two in the median interspaces, and the largest one extending from the lower median venule to the submedian; the cell well washed with brownish ochre, but not containing pale yellow rays as in *Pontiac.* The hind wings contain no bands or spots, but the area between the median vein and its lower branch and the submedian is paler than the ground color, and there is another pale ray from the middle of the cell outward to near the outer margin.

Female.—Of the same shade of brown as the male, both having a slight vinous reflection. The fore wings have a small yellow spot at the end of the cell, and the usual outer row of spots. These consist of the three anteapical in the subcostal interspaces, the first one only a few scales, the others elongate; the fourth and fifth at the end of the cell, the first only a few scales; the sixth and seventh in the median interspaces, rounded inwardly, concave externally; the eighth in the medio-submedian space, partially or wholly divided in the middle. The hind wings are similar to those of the male, the stripe from the middle of the cell to near the outer margin quite prominent, but mere traces of three other elongate spots. Under side as in the male.

Body concolorous with the wings above, beneath pale yellow.

Larva and food-plant unknown.

Nebraska ; Whitings, Ind.

163. PAMPHILA ARPA, Bd.—Lec.

Expanse of wings from 1.6 to 1.8 inches.

Male.—Upper surface dark olivaceous brown, the base of the fore wings and along the costa with yellowish scales. The cell of the fore wings, a patch before the stigma, and a band outside the stigma, crossing three intervenular spaces and a little less than half the distance to the outer margin, are clear golden yellow. There are also five or six yellow rays between the subcostal venules. Stigma long, slender, somewhat constricted in the middle, oblique, black. The hind wings are sprinkled a little with yellow scales in the middle, the inner part with olivaceous hairs.

Under side dark golden-yellow, the posterior part of the fore wings blackish. Across the outer portion of the fore wings are about three pale spots, there being scarcely a trace of one on the submedian vein. Hind wings without spots.

Female.—The upper surface the same brown as the male, with a little sprinkling of yellow scales at the base. The fore wings have the rays between the subcostal venules, but not so distinct, except the lower two; and there are three spots in an oblique row,—one on the submedian vein and two between the branches of the median. Under side as in the male.

The mature larva, according to Dr. Chapman, is nearly two inches long; pale green striped with yellow, the joints after the second thickly lined with fine streaks of green and yellow. Spiracles black. Head high, narrow, blackish, bordered round the top and sides by white, and with two white incurved (concave to each other) streaks on the upper third of the face; these separated by velvety black.

Chrysalis.—Length 1.2 inches, nearly cylindrical, light brown, covered with white powder; the abdominal joints pubescent; the wing-cases prolonged into a short subulate point; the abdomen long, tapering slightly, the end bluntly rounded.

The larva feeds on saw-palmetto, forming a tube of the fan-like segments of the leaves, in which it lies concealed until it changes.

Gulf States.

164. Pamphila Palatka, Edw.

Expanse of wings from 1.45 to 1.5 inches.

Male.—Upper surface dark brown; the outer third of the cell of the fore wings, and about the middle third of the wing below the cell, except a narrow posterior border, clear yellow; the basal third of the wing washed with yellow, blending into the clear yellow of the middle, so that without a glass the whole of this appears yellow. Beyond the cell there are about three yellow spots in the subcostal interspaces, sometimes the first and second obscure. Stigma oblique, narrow, broken in the middle, dull black. There is also a bar across the end of the cell. Hind wings yellow, with a broad terminal and costal border.

Under side of hind wings and anterior part of fore wings brown, heavily overlaid with russety scales, so as to give these parts a russety brown appearance; the posterior part of the fore wings blackish. The yellow of the fore wings is repeated, that part in the cell tinged with orange and expanded basally. The hind wings are without distinct spots, but have a ray through the cell a little paler than the rest of the wing.

Female.—Resembles the male, but lacks the sex-mark above, and has the anteapical row of spots a little more distinct. The under side has the costa of the fore wings more suffused with orange.

Body dark brown above, sprinkled with yellow scales, and with greenish-yellow hairs. Under side of thorax pale yellowish; abdomen buff, tinged with brown; palpi pale yellow, brownish at the ends. Antennæ yellow beneath, the tips fulvous; above annulate with brown and yellow; the knob brown.

The mature larva is two inches long, cylindrical, with the collar a black line connecting two black lateral spots. Anal plate semicircular, projecting. Body yellowish green, thickly dotted with minute, dark, hair-tipped tubercles; spiracles black; under side bluish. Head obliquely projecting, brownish, the upper part of the face white and marked by three black stripes.

According to Dr. Chapman, the larva feeds on a species of grass (*Claudium effusum*), drawing the faces of the strongly-keeled leaves together, and in the tube thus formed lying concealed when not feeding.

Gulf States, Nebraska.

165. Pamphila Vitellius, Sm.—Abb.

Expanse of wings 1.2 inches.

Male.—Upper surface bright yellow; fore wings with

an outer border of dark bronzy brown, in width about one-fifth the length of the wing, and extending as a narrower border along the hind margin, where it is sprinkled with yellow scales. Costa narrowly black. Hind wings with the margin bordered with brown, leaving the middle area yellow, suffused somewhat with brownish; hairs yellow. Fringes pale yellowish, dusky at apex of fore wings.

Under side clear rich yellow, without spots, the posterior part of the fore wings smoky black. Body covered with dusky yellow hairs above, paler yellow beneath.

Female.—Outer fourth of the fore wing, and a border along the hind margin of about the same width, the same dark bronzy brown as the border of the male; the remaining area of the fore wings dull yellow sprinkled with brown, also brown along the veins. Hind wings the same brown as the border of the fore wings, with the centre slightly washed with yellow. Under side as in the male.

Body dark brown above, with yellow hairs; under side pale yellow, the palpi and the under side of the antennæ paler than the body.

Larva and food-plant unknown.

Georgia, Southern Texas, Iowa, Nebraska.

166. Pamphila Delaware, Edw.

Expanse of wings 1.2 inches.

Male.—Upper surface yellow, slightly fulvous-tinted; the veins, except the subcostal, brown; the fore wings with a dark brown outer border, in width about one-eighth the length of the wing, extending only a little along the hind margin, its inner edge but slightly crenate.

Hind wings with the outer border about the same width, but the costal and inner margin a little broader.

Under side about the same color as the upper, without spots; the basal half of the posterior part of the fore wings smoky black, extending as a narrow posterior border along the outer half.

Female.—Dark brown marked with pale yellow. The basal two fifths of the fore wings brown; beyond this a band of yellow extending half-way to outer margin, and not quite reaching either margin, separated into spots by the veins and venules, making in all ten intervenular yellow spots, besides the one that fills the outer third of the cell. Hind wings as in the males, but paler, the veins dark.

Under side almost a lemon-yellow, the black on the fore wings as in the males, with the addition of a border from the posterior angle half-way to the apex, and most of the outline of the cell black.

The larva, according to Dr. Chapman, feeds upon a large species of grass (*Erianthus alopecuroides*), rolling itself in a leaf. When full grown it is one inch long, fusiform, bluish white; collar black, ending in a dot on each side; a lunate black band on joint 13 and anal plate. The surface is thickly dotted with minute black tubercles. Head oval, oblique, white, smooth, slightly bilobed; a dark band about the top and sides, a black vertical streak on the middle of the face, and a short streak of the same color on each side of this.

The chrysalis is narrow, greenish white; the headcase blunt, black-tubercled, and bristly; the last joint black.

Massachusetts to Florida, Illinois, Kansas, Montana.

167. PAMPHILA BYSSUS, Edw.

Expanse of wings from 1.6 to 1.7 inches.

Male.—Upper surface dark glossy brown; basal half of costa of fore wings ferruginous, as well as a little of the cell below adjoining; at the end of the cell an irregular yellow-fulvous bar within; across the disk a bent yellow-fulvous band, starting on costal margin about three-fifths the distance from base to apex, bending round the cell, and continued to middle of submedian vein, narrow at top, but below the cell widening rapidly, on the submedian being in width about one-third the length of this part of the wing. The basal half of hind margin washed with fulvous.

The hind wings have a fulvous patch in the middle, consisting of a broad band beyond the cell, not reaching either margin, and a small spot in the cell, sometimes the spot obsolete. The hairs of basal area next the inner margin fulvous. Fringes of fore wings dark gray-brown, of hind wings lighter.

Under side wholly ferruginous (individuals varying a little in shade), except that the posterior half of the fore wings is blackish. The spots of the upper side are repeated indistinctly; on the hind wings, in most examples, the surface is without spots, in some there is a faint paler color indicating the patch of the upper side. The veins and branches are a shade more yellow than the ground color.

Body above covered with fulvous hairs; beneath, the thorax and ventral part of abdomen yellowish, sides of abdomen and legs ferruginous.

Female.—Upper side of same brown as the male, and

marked with fulvous in a similar manner, but the band is narrow and of nearly uniform width throughout, except at the bend opposite the cell, where it is much restricted.

Under side as in the male, but in six cases out of seven the band of the upper side of the hind wings is indicated below with much distinctness.

In one male the fulvous band is diffused, and the basal area is also fulvous, so that the whole of the wing is of that color, except a stripe around the end of the cell, and the outer margin. The males have no stigma.

Larva and food-plant unknown.

Indian River, Florida; Texas.

168. PAMPHILA OSYKA, Edw.

Expanse of wings 1.1 inches.

Male.—Upper surface uniform brown, with a slight green tinge, without spots. Stigma long, broadest at the upper part, depressed in the middle. Under side brown. Body gray beneath; palpi greenish white.

Female.—Of the same size as the male, and the same color above except the stigma. Under side clear gray, except on hind margin of fore wings, which is brown. On the costa of fore wings are three minute semi-transparent spots in a line, and on the disk are two others at an obtuse angle with the first. Palpi white.

Larva and food-plant unknown.

Gulf States; Whitings, Ind.

169. PAMPHILA EUFALA, Edw.

Expanse of wings from 1.1 to 1.2 inches.

Male.—Upper surface dark olive-brown. The fore

wings have three small white semi-transparent spots beyond the cell in the subcostal interspaces, and two spots
below in the submedian interspaces, the first subquadrate, and the second oblong, about twice as large as the
first; both small. Besides these there are one or two
opaque white points at the end of the cell; wanting in
some examples. Hind wings without spots.

Under side a little paler than the upper, hind wings
and costal and apical portions of fore wings sprinkled
with gray. The spots of the fore wings are repeated.

Female.—Similar to the male, but on both surfaces
there is a slight violet tint, more apparent along the
veins.

Body black above, hairs dark olivaceous brown;
under side, and palpi, whitish.

Larva and food-plant unknown.

Florida, Texas.

170. PAMPHILA FUSCA, Gr.—Rob.

Expanse of wings 1.05 inches.

Male.—Upper surface glossy olivaceous blackish,
without spots, but with a faint and variable yellowish-brown reflection equally distributed. Fringes pale,
without spots.

Under side of wings shaded with lustrous golden-
brown scales. Fore wings about the same shade as
above, with the golden-brown on the anterior and terminal portions. Hind wings evenly covered with pale
golden-brown scales, except a space before the inner
margin, extending from the base to the outer margin.

Body above concolorous with the wings; beneath pale
yellowish olive, palpi yellowish white.

Female.—Similar to the male, but the palpi are paler, as is also the under side of the abdomen.

Larva and food-plant not known.

Gulf States.

171. PAMPHILA HIANNA, Scud.

Expanse of wings from 1.3 to 1.45 inches.

Male.—Upper surface dark brown; the fore wings with three small white anteapical spots in the subcostal interspaces, one larger in the first median interspace, and a small one in the upper part of the outer end of the cell. Hind wings without spots.

Under side about the same color as above, the basal two thirds, except posterior part of fore wings, sprinkled with brown, the outer part sprinkled with gray. The spots of the upper side are reproduced with a little more distinctness, and there is a dim curved line in the second median interspace of the fore wings. Hind wings with a small white point below the costa beyond the middle.

Female.—Differs from the male in having the spots larger, two small ones opposite the cell of the fore wings, out of line with the others, and a spot somewhat larger than the others in the base of the second median interspace. On the under side the hind wings have a small spot on the anterior part, near the base.

Body blackish above and below; palpi dark gray.

The larva and food-plant are not known.

Massachusetts to Nebraska.

172. PAMPHILA VIATOR, Edw.

Expanse of wings 1.5 inches.

Male.—Upper side of fore wings dark brown, with a

reddish tint. There is a double yellow spot in the outer
end of the cell, and a discal row of spots across the wing.
The first three of these are in the subcostal interspaces,
the first subobsolete or obsolete; below these are three
more,—two in the median interspaces, and one irregular
one, somewhat hour-glass-shaped, with the lower part the
broadest, extends from the submedian to the lower branch
of the median. This may be divided in the middle into
two spots.

The hind wings have a broad brown margin, broader
along the costa and outer part than along the inner margin, the whole of the rest of the wing yellow, broken into
long spots by the brown veins.

Under side of fore wings smoky along the hind margin, reddish brown along the costa and apex; the spots
of the upper side repeated indistinctly. The hind wings
pale brown, with the spots repeated indistinctly.

Some examples have two small spots beyond the cell
of the fore wings, and the other spots somewhat enlarged,
the spot on the submedian with the lower part shading
out towards the base. The two spots beyond the cell do
not show on the under side.

Female similar to the male, but a little larger.

Body above brown, below gray; palpi whitish; club
of antennæ brown tipped with fulvous.

Larva and food-plant unknown.

Gulf States, Massachusetts, New Jersey, Illinois, Wisconsin.

173. AMBLYSCIRTES VIALIS, Edw.

Expanse of wings 1 inch.

Upper surface dark blackish brown, with slight violet

reflection. The fore wings have three small white ante-apical dots in the subcostal interspaces, about three-fourths the distance from the base to the apex, and traces of spots in the median interspaces. Hind wings without spots. Fringes gray, spotted with dark brown at the ends of the veins.

Under side as above; the apical half of the fore wings, and all of the hind wings except a ray before the anal angle, washed with lilac scales, the anterior basal part of the hind wings only sprinkled. There is a clustering of the gray scales across the disk of the hind wings, constituting a rather indistinct connected series of about four spots forming a curve, made more apparent by there being less gray just before these than on other portions of the wing. There is a white spot in the fringe at the apex of each wing.

Body olivaceous brown above, lilac-gray below, including the palpi.

In the larval state this species feeds on grasses, the butterfly appearing from the first to the middle of July.

Orono, Maine; Middle, Southern, and Western States.

174. AMBLYSCIRTES Eos, Edw.

Expanse of wings 1 inch.

Upper surface grayish brown. The fore wings have three white spots in line from the costa back, as in *A. Vialis*, but no other spots. Fringes alternate white and fuscous on the fore wings, but on the hind wings fuscous only at the tips of three or four venules in the middle of the margin.

Under side brown, with a whitish or chalky tint at

the apex of the fore wings and along the outer margin,
and over most of the hind wings, quite dense on the outer
margin. The spots on the fore wings are repeated, a
little enlarged, and accompanied by a fourth below the
others and towards the outer margin. The hind wings
have a mesial row of whitish points, not reaching either
margin, irregular, rather forming a double row, with a
similar point in the cell and two in the interspace above
the cell.

Texas, Georgia, Florida.

175. AMBLYSCIRTES SAMOSET, Scud.

Expanse of wings 1.1 inches.

Upper surface dark brown, sprinkled with greenish
scales, which give a greenish shade to the wings. The
fore wings have a row of pale yellow spots beyond the
middle, consisting of three near the costa in the subcostal
interspaces, mere dots; the fourth, in line with these, in
the second interspace beyond the cell; the fifth and sixth
are in the median interspaces, the sixth much larger than
any of the others; the seventh some light scales above
the submedian vein, forming two indistinct spots. There
are traces of a small spot in the end of the cell. The
hind wings have a faint trace of a discal line.

Under side lighter than the upper, overlaid with greenish scales, with the exception of the posterior part of the
fore wings. The spots of the fore wings are repeated,
with two at the end of the cell. Hind wings with the
discal band distinct, but not reaching either margin, an
indistinct spot in the cell and two above it. The discal
band is composed of about five white spots which are
united. The fringes are white, marked with dark brown

at the ends of the veins on the fore wings and the middle
veins of the hind wings.

The larva is said to feed on a coarse grass (*Andropo-
gon*). The butterfly is on the wing through the middle
of June.

Northern and Middle States, Maine to Georgia, Wisconsin, Iowa.

176. AMBLYSCIRTES TEXTOR, Hüb.

Expanse of wings 1.15 inches.

Upper surface olive-brown, the fore wings with an
irregular row of six small angular white spots running
from the costa back to the middle of the wing, about
three-fourths the length of the wing from the base. The
first three of these are in the usual subcostal interspaces,
the next two in the interspaces beyond the cell, the sixth
in the first median interspace. There is a trace of the
seventh in the upper part of the second median interspace, farther back than the others. The first is a mere
point, but there is a gradual increase in size up to the
third and fourth, this being out of line with the others,
the fourth, fifth, and sixth about the same size. The
hind wings are without spots, but there is a faint trace
of the discal band of the under side. Fringes long,
whitish, cut with brown at the ends of the veins.

Under side of fore wings as above, with the spots
more distinct, the seventh being an elongate spot reaching from vein to vein, and a small spot above the submedian. In addition to these there are two minute dots
at the end of the cell. All these are yellowish white.

Hind wings brown, with a gray-violet tint, except
near the inner margin. Across the disk, beyond the

cell, is a tortuous connected row of irregular yellowish-white spots extending from costa near apex to near inner margin. This is somewhat dilated at the extremities, and sends a short ray outward beyond the cell. Across the end of the cell extends another somewhat broken row, consisting of two or three spots in and below the cell, the two being connected by the white veins, giving the hind wings a somewhat reticulated appearance.

Food-plant not known.

North Carolina to Texas.

SECTION II.

In this division the knob of the antennæ is spindle-shaped. The larvæ are more or less cylindrical, with the head usually larger than the second segment. Habits of larva and pupa mostly similar to those of Section I.

177. PYRGUS TESSELLATA, Scud.

Expanse of wings 1.2 inches.

Male.—Upper surface black, the basal third and hind margin of fore wings overlaid with white hairs, as also the inner part of hind wings. The outer two thirds of fore wings contain about thirty white spots arranged somewhat in four irregular transverse rows; and five more on the costal edge, as shown in Fig. 86. The hind wings have about eighteen spots, arranged in three rows, the spots of the inner row the largest, the middle ones crescents, the outer ones points. Fringes white, with black at the ends of the veins.

FIG. 86.

Pyrgus Tessellata, male (natural size).

Under side of fore wings yellowish white along the costa and the outer margin, the spots of the upper surface repeated, but more or less blended, the space between the spots brownish black. Hind wings white, faintly yellowish; a spot near the base, two irregular transverse bands, and a marginal row of lunules are brownish, these corresponding to the spaces between the rows of white above.

Female.—Darker, owing to the spots being smaller, the marginal row of points subobsolete. Under side also darker.

Pennsylvania to the Gulf, Atlantic to the Pacific.

178. Pyrgus Centaureæ, Ramb.

Expanse of wings 1.25 inches.

Upper surface black, tinged with brown, sprinkled somewhat with white scales over the basal half. There is a bar of white in the end of the cell of the fore wings, a less distinct spot of the same below the middle of the cell, and a subterminal row of white spots. There are first three spots in the subcostal interspaces three-fourths of the distance from the base to the apex; two spots beyond the cell, the upper half-way from the third spot to the margin; spot five in line with the first three; and an oblique row of four more,—two in the median interspaces and the other two in the medio-submedian interspace, one next to each vein. Besides this, the cross-vein at the end of the cell has some white scales. In addition to these, there are five white spots on the costal edge.

The hind wings have two obscure rows of white spots, the first crossing the end of the cell, the spot on the cell being the most distinct; the second subterminal, consisting of a series of intervenular, somewhat sagittate spots. Fringes white, cut with black at the ends of the veins.

Under side a little paler than the upper, sprinkled over with white scales so as to be grayish brown. The spots of the fore wings are repeated, but enlarged and coalescing. The hind wings have three distinct bands of white, one near the base which does not show on the upper side, consisting of three patches united; the middle one enlarged and blended; the subterminal one not much more prominent than on the upper surface.

Body black above, with gray hairs, the scales and hairs below white; antennæ annulate with black and white; tip of club ferruginous.

New York, West Virginia, North Carolina, Colorado.

179. NISONIADES BRIZO, Bd.—Lec.

Expanse of wings from 1.3 to 1.5 inches.

Upper surface dark brown, the outer portion of the fore wings sprinkled with gray. Near the outer margin of the fore wings is a row of small gray spots, and between these and the cell is a row of larger contiguous gray spots, varying from oval to orbicular, bordered, except where they come together, by a line of darker brown than the ground color; the spots of the upper end of the row have the brown pointed outwardly. A similar row crosses the wing about through the middle, but this has no points on the outside; and there is a dark spot at the base of the cell. The hind wings have two wavy rows of ochre-yellow spots, which are dull and obscure.

Under side a little paler than the upper, with two rows of white spots parallel with the outer margin, common to both wings; the hind margin of the fore wings dull whitish.

Atlantic States to the Rocky Mountains, Colorado, Arizona.

180. NISONIADES ICELUS, Lintn.

Expanse of wings from 1.15 to 1.25 inches.

Upper surface blackish brown, sprinkled with pale bluish scales. A band, somewhat lighter than the basal part of the wing, crosses the fore wings so as to bring the end of the cell in its middle. This band is heavily overlaid with bluish scales, especially on the costal half. Preceding this is a band of darker brown, subquadrate, contiguous spots, with a slight sprinkling of bluish scales. Beyond the band which crosses the end of the cell is another band of ovate spots, the bases rounded, the costal half with the spots narrower and more pointed, the anterior four or five with whitish, imperfectly-defined spots in their centres, and the rest overlaid with bluish scales. Between this band and the outer margin is a row of obscure brown spots without the pale scales. The hind wings have two irregular rows of dull ochraceous spots along the outer margin.

Under side paler than the upper side, the hind wings with the two rows of spots repeated. The fore wings have two rows of spots parallel with the outer margin, both elongate, the outer on the edge and extending into the fringe. Inside these, and corresponding with the second band of the upper side, is a row of white spots, the first six small, the seventh and eighth larger and quadrate, the ninth and tenth obscure.

According to Mr. Lintner, the egg is of a pale green

color, in shape semi-ellipsoidal, its base flat, and its apex depressed between the tips of the ribs, which terminate exterior to the depression. It is distinctly ribbed with from ten to twelve longitudinal ribs, and connecting the ribs are from thirty to thirty-five transverse striæ. Diameter, .031 of an inch ; height, .028 of an inch.

The larva is unknown. The butterfly may be seen in June.

New England to Michigan, Illinois, Florida, Colorado, Arizona, Washington Territory.

181. NISONIADES SOMNUS, Lintn.

Expanse of wings from 1.3 to 1.45 inches.

Male.—Dark brown in color, approaching *N. Persius.* Fore wings without the anteapical white spots above, and the large patch of bluish-white scales resting on the discal cross-vein of *N. Icelus.* The black transverse bands have the position and character of those of *N. Icelus*, but are almost lost in the ground color. Hind wings nearly as dark as the fore wings, showing indistinctly the two rows of pale brown spots.

Under side of wings bronze by reflection. The fore wings have a short costo-apical white streak in cell 8, or between the third and fourth subcostal venules, and a minute white dot above this, or in cell 9 (*N. Icelus* usually has a subquadrangular white spot in the upper interspace beyond the cell and the two lower subcostal interspaces, and occasionally the spots form a continuous line nearly across the wing from the lower median interspace to next the lower subcostal interspace). There is an intervenular series of pale streaks, and on the hind wings the two rows of yellow-brown spots are indistinct.

Female.—Paler brown than the male. The two transverse bands of the fore wings are quite distinct, and between them, on the discal cross-vein, is a conspicuous patch of whitish scales; no white anteapical spots. Upon the margin is a row of rounded brown spots, separated from the contiguous band by whitish scales. The bands are broader than in *N. Icelus*, and are almost drawn together on the second submedian vein; the connected series of spots composing each are shaped much as in *N. Icelus*, are heavily bordered with black, and bear bluish scales. The hind wings contain a geminate discal mark, a submarginal row of yellowish spots much bent inwardly opposite the cell, and a marginal row of small linear whitish spots.

Under side bronze like the male. The hind wings have the two rows of spots repeated; the fore wings have a marginal row of linear whitish spots, a regularly curved submarginal row of eight oblong yellowish spots, and a single white anteapical spot.

The palpi are shorter than in *N. Icelus*, shaggy, some of the hairs of the second joint extending to the tip of the third. The tibiæ of the posterior pair of legs are without the pencil of hairs characteristic of *N. Icelus*.

Indian River, Florida.

182. NISONIADES LUCILIUS, Lintn.

Expanse of wings from 1.1 to 1.25 inches.

Upper surface dark brown, with a red lustrous reflection. Like *N. Icelus*, there is a paler band at the end of the cell of the fore wings, and this is preceded and followed by a darker band, the inner not well defined, and interrupted. In the female, and sometimes in the

male, there is a white hyaline spot resting on the **outer
end** of the two cellular teeth formed by this band, sometimes obsolete. **The** submarginal band consists of interspaceal, sagittate, fuscous spots, which are somewhat
squarely truncated anteriorly and have umber-colored
scales centrally. Its course is direct from the submedian
vein to the subcostal venule, whence it is broadly reflected
anteriorly to the costal margin, embracing in this portion
four interspaceal minute white hyaline spots, of **which
the first, third, and fourth are nearly in line, the second
and largest lying behind.** Between the median venules
are two hyaline spots, of which the inner one is sometimes obsolete in the male or wholly absent. Between
these two bands the ground color is umber-brown, with
a few bluish-gray scales towards the submarginal band,
and a large number between the subcostal venules. The
sagittate spots of the submarginal band are bordered
externally with gray, followed by a series of round umber spots having a few gray scales resting on obscure
yellowish spots between them and the brown marginal
line. Fringes umber, with a very few basal **gray scales.**

Hind wings of a **more uniform brown** than the fore
wings, **and more shaded with red, with two rows of obscure yellowish-brown marginal spots;** the discal spots
at the end of the cell barely seen.

Under side **reddish** brown, the fore wings conspicuously so at the apex. The four subapical hyaline spots
are constant, and there is usually a small white spot in
the cell. The median spots are larger than the subapical,
and are subquadrangular in form. Hind wings without
discal spots. Both wings with **two rows** of spots along
the outer margin.

The eggs are .03 of an inch in diameter, marked with fourteen ribs and twenty-five transverse striæ.

The larva at maturity is .8 of an inch long, yellowish green in color, with a bluish-green dorsal line. The legs are tipped with fuscous. On joint 8 is an oblong yellow spot on each side of the dorsal line, a similar mark on joint 2, and a brown spot on the head.

The chrysalis is cylindrical, conical, not angulated, thorax slightly elevated. Head-case rounded in front, depressed below a line drawn from the anal spine across the base of the wings to the humeral tubercle. Towards the close of this period the eye-cases are purple, the wing-cases whitish, the abdomen green, except at the tip, where it is brown.

Food-plant *Aquilegia Canadensis.* There are two broods in a season, possibly three, the butterflies appearing in May and about the last of August or in the fore part of September.

New York, Middle and Western States.

183. Nisoniades Persius, Scud.

Expanse of wings from 1.2 to 1.4 inches.

Upper surface blackish brown, the outer part of the fore wings sprinkled with pale bluish scales in the males, but in the females a little at the base also. Like the other species, this has a mesial band crossing the discal cell, and a submarginal band, the first more obscure than the second, except below the median vein and in the cell; the upper point in the cell containing a distinct hyaline spot in the females, but more obscure in the males. In the outer band spots 1 and 4 and 7 and 8 contain each a distinct white hyaline spot, the second

larger and a little out of line; spots 9 and 10 are united
into an hour-glass-shaped spot. Spots 1 to 5 are narrowly ellipsoidal, 6 to 8 are somewhat truncated internally, all the others are strongly pointed internally. Between these two bands the space in the end of the cell
and between the median and submedian veins is lighter
and more reddish brown than the rest of the wing. Between the submarginal band and the margin is a series
of roundish brown spots, each tipped externally with a
gray crescentic spot resting on a terminal brown line.

Hind wings more of a reddish brown than the fore
wings, with two rows of indistinct ochraceous spots near
the outer margin, and a spot at the end of the cell. The
males are darker brown than the females, and for that
reason the bands are more obscure.

Under side of the female grayish brown, the apical
portion of the fore wings gray; the white hyaline spots
are repeated, and both wings have two marginal rows
of whitish spots. The males, as above, are darker than
the females.

According to Mr. S. H. Scudder, the larva feeds on
willow, poplar, and *Lespedeza capitata*. The butterflies
are to be seen in June.

New England to Montana, Washington Territory,
Colorado, New Mexico.

184. NISONIADES AUSONIUS, Lintn.

Expanse of wings 1.06 inches.

Fore wings above pale umber-brown, with grayish
scales sprinkled over most of the surface (more diffused
than in the other species), except on the fuscous bands,
showing especially beyond the submarginal band. There

are two brown basilar spots resting on the subcostal and median veins, not so dark as those of the disk. The discal band, usually continuous in this genus, here consists of three elongate fuscous dashes (appearing to the unaided eye as a single spot) resting on the subcostal near the discal cross-vein, extending nearly half-way to the median, the intervening space having merely an indication of the spot, which appears distinctly in most of the species as the inner cellular tooth of the discal band; following this is an obscure fuscous spot at the fork of the first and second median venules, and, beyond, the usual hour-glass-shaped spot, extending from the second median venule to the submedian, with its constriction on the interspaceal fold. The discal cross-vein is conspicuously marked in brown. The submarginal band of fuscous spots is doubly curved, being convex towards the outer margin from the costa to the third median venule, thence concave to its termination at the submedian. It consists of four acutely ellipsoidal fuscous spots between the subcostal venules, which are wholly destitute of the usual hyaline spots, followed by three others of similar form but of greater breadth, the next subacute externally, and the last similar in outline to the corresponding one of the discal band. There is a marginal row of interspaceal brown spots, the first four of which are surrounded by gray scales and lie near the margin, and the remaining four more remote than in *N. Martialis;* also a row of obscure brown spots resting on the tips of the venules and extending on the fringe.

Hind wings of a darker ground than the fore wings, sprinkled with blackish scales, darker basally, and with pale yellow-brown spots. Discal spot and marginal row obsolete, submarginal crescentic in form. On each side of the submarginal row of spots is a range of oval fuscous spots, subconnected.

Under side reddish brown, with the terminal margin gray. The fore wings have the fuscous spots of the submarginal band and marginal row as on the upper surface; of the discal band the spots in the cell are alone obscurely visible; the basal ones are lost in the general color. The marginal brown spots below the subcostal venules rest centrally on elliptical gray patches, while those of the hind wings approach a semioval form, and are preceded by gray crescents, which nearly enclose them by uniting with some marginal gray scales; at the tips of these crescents a submarginal row of fuscous spots is obscurely seen. Discal spot obsolete.

Middle States, West Virginia.

185. NISONIADES MARTIALIS, Scud.

Expanse of wings 1.5 inches.

Upper surface of female blackish towards the tip of the fore wings, the rest slightly grayish. Near the base of the cell a dark blackish-brown spot, and beyond this the mesial and submarginal bands of spots of the same color, the first obscure on the costa and broken on the median vein, the last spot being placed farther towards the base, so as to be out of line with the others. On the upper outer point in the cell is a somewhat elongate white hyaline spot, expanded externally and posteriorly, and on the lower point traces of a second spot. Submarginal row as in *N. Persius*, but the hyaline spots much larger, the one in spot 8 being twice as long as broad, and the hour-glass spots above the submedian vein have pale brown centres. The marginal row of brown spots lacks the gray crescents, there being only mere traces of them, and are not set in a grayish field, there being some bluish scales inside this row over most of the wing, but none outside; and there is not so much difference between the color of the end of the cell and the rest of the wing as in *N. Persius.* Hind wing with only traces of spots in the usual places.

Under side a little paler brown than the upper, all the hyaline spots distinct. The marginal and submarginal bands of the fore wings are dimly outlined, the outer set in a paler ground. The usual double row of spots on the hind wings is dim.

The male differs from the female in having the bands and hyaline spots less distinct, and the ground color a little darker.

Atlantic States to Florida, Mississippi Valley, Kansas, Colorado.

186. NISONIADES JUVENALIS, Fab.

Expanse of wings from 1.3 to 1.6 inches.

Upper surface dark smoky brown, so dark that the usual bands are a little obscure. These are the mesial and submarginal bands of spots on the fore wings, and an obscure spot at the base of the cell. The mesial band would be obscure were it not for the few gray scales that border the spots. The upper point in the cell contains a small hyaline spot. The submarginal band contains five or six hyaline spots, the first four situated in spots 1 to 4 of the band, and the fifth in spot 7. The spots of this band are shorter and less pointed than in *N. Persius* and *N. Martialis,* the last two scarcely united;

a few gray scales each side of the band. Outside this band is a series of blackish-brown roundish spots, with a few whitish scales between each spot and the obscure marginal line.

Hind wings dark smoky brown, with a mere trace of the usual double row of marginal spots.

Under side nearly as dark as the upper, with purple reflections, especially on the hind wings. The hind wings show mere traces of the usual two rows of spots; the fore wings have the hyaline spots distinct, and the marginal and submarginal bands are to be seen, but the field on which they rest is not so pale as in *N. Martialis.* Fringes concolorous with the wings.

According to Harris, the larva of this species feeds on species of Apios and Lathyrus, and perhaps other Leguminosæ. It is green, with pale stripes, and has a heart-shaped brown head.

The chrysalis is rather long and tapering, pale yellowish brown, with a few minute hairs on the body, and with the tongue-case prominent and projecting beyond the middle of the breast. There are two broods of these insects, the last hibernating in the chrysalis state. Probably there are more than two broods in the Southern States.

Atlantic States to Florida, Mississippi Valley, Colorado, Arizona.

187. NISONIADES PETRONIUS, Lintn.

Expanse of wings from 1.9 to 2 inches.

Upper surface dark blackish brown, the submarginal band of subsagittate spots oblique, black, being more drawn out inwardly towards the base as it approaches the hind margin. The black markings of the wings are

more strongly contrasted with the dark brown ground
than in *N. Nævius*, but less so than in *N. Juvenalis,*—
about equal to *N. Persius.* The white hyaline spots of
the fore wings are of medium size, smaller than in the
average *N. Juvenalis;* the spot in the discal cell is small;
that in the upper median interspace on the transverse
band of sagittate spots is crescentic, concave towards the
base; below this, in the second median interspace, is a
smaller white spot, wanting in some examples. The
four anteapical spots in the same band are not quite
in a line, as they are in *N. Juvenalis*, the third standing a little farther towards the base than the others;
the second and third being oblong, instead of quadrate
as in *N. Juvenalis.*

There are fewer white scales on the fore wings than
in *N. Juvenalis*, there being scarcely any on the basal
side of the submarginal band, and only a few on the
outer side. The pale on the outside of the marginal
band is very dim. The usual two submarginal rows on
the hind wings are very obscure.

Under side reddish brown, especially the hind wings,
which show two rows of pale brown submarginal spots,
which become obsolete before reaching the front margin
of the wing, and wholly want the white spots in cells 6
and 7 which characterize *N. Juvenalis.* The white spots
of the fore wings are larger than above.

Head: above the eyes and just behind the "locklet"
are a few white scales; behind and beneath the eyes are
some pale yellow-brown scales, and similar-colored hairs
compose most of the palpal covering, in strong contrast
with the dark brown color of the legs, thorax, and
abdomen.

This is the largest species of the genus, and is separable from *Juvenalis* and *Propertius* by its darker color, less distinct ornamentation, less rounded wings, and the absence of white spots of the hind wings in cells 6 and 7 ; from *Nœvius* by its larger size, more distinct markings, and the contrasting lighter shade of the palpi.

Indian River, Florida.

188. Nisoniades Nævius, Lintn.

Expanse of wings from 1.45 to 1.65 inches.

Upper surface fuscous, almost black, with a purple reflection. The fore wings have four minute, subquadrangular, costo-apical, hyaline spots, of which the fourth may be obsolete, resting on the first four spots of the submarginal band, and a similar spot on spot 7 of this band, but none on spot 8 nor at the end of the discal cell. An irregular umber-brown spot centres on the discal cross-vein, and between the median and submedian veins is another, showing more distinctly in the female. The terminal row of obscure, rounded, intervenular fuscous spots rest on a dark umber-brown ground. All the markings are nearly lost in the dark ground ; those best defined are two confluent trapezoidal spots above the submedian vein, forming the posterior termination of the transverse row of spots, and defined without and within by a W in umber-brown. The spots of the transverse row are not of the ordinary sagittate form. The hind wings dark brown, showing faintly the two rows of intervenular paler brown spots, more distinctly in the female. Fringes dark brown, lighter upon their outer half in some males, and pale, approaching whitish, in the female.

Under side paler brown, and showing more or less distinctly the two ordinary rows of pale brown spots towards the outer margin, and in one male a white spot in the cell, not seen on the upper surface. Head and palpi concolorous with the thorax, abdomen, and legs.

Indian River, Florida.

189. PHOLISORA CATULLUS, Fab.

Expanse of wings from 1 inch to 1.1 inches.

Upper surface deep brownish black. The fore wings have a submarginal row of eight white hyaline spots, 3, 4, and 5 curving outward beyond the cell, 6 and 7 in the two median interspaces, 8 above the submedian. The first three are subquadrate, the rest mere dots. There is also a small spot at the end of the cell. Hind wings without marks. Fringes concolorous with the wings.

Under side more of a brownish black, the spots on the fore wings repeated. The body and head are black above, the head with three longitudinal white stripes, the ends of the palpal hairs the same color; the palpi and head white beneath, the body black.

Fig. 87 represents the egg of this species. The larva feeds on *Monarda punctata*, *Chenopodium album*, Ambrosia, and perhaps some related plants. It is found in the United States generally.

FIG. 87.

Egg of P. Catullus, × 15.

190. PHOLISORA HAYHURSTII, Edw.

Expanse of wings from 1 inch to 1.1 inches.

Upper surface blackish brown, both wings crossed by two deeper brown bands, more distinct in the female;

the mesial one crossing the fore wings before the end of
the cell, the other submarginal, just beyond the end of
the cell. In the submarginal band the fore wings have
two small white spots near the costa, and a white point
in the second median interspace. Hind wings dentate
on the outer margin, the fringe of the points concolorous
with the wing, that of the hollows paler. The bands
show more plainly on the female than on the male, on
account of the ground color being a little paler, and there
is a trace of a third anteapical spot.

Under side a little paler than the upper, a little sprinkled with ochraceous scales, the spots of the upper side of
the wings repeated.

Body above concolorous with the wings, below grayish ; head and palpi above with a few brownish-yellow
scales, below white.

West Virginia to Kansas, Florida, Texas, New
Mexico.

191. EUDAMUS PYLADES, Scud.

Expanse of wings from 1.4 to 1.5 inches.

Upper surface dark brown. The fore wings are
marked by several white hyaline spots : first three anteapical in the subcostal interspaces ; three more above the
cell, about two-fifths the distance from the base of the
wing to the apex,—both of these in a line back from the
costa ; two more in the median interspaces, and one below
the lower median venule, close to this venule and nearer
the margin than the others. All of these are small,
those of the costal region subquadrate, the others triangular, the one in the lower median interspace more or
less obsolete. There is also a curved brown mark be-

yond the cell from the lower end of the anteapical row. The lower three are arranged in the form of a triangle, and when one is obsolete its place is indicated by a mark of deeper brown than the rest of the wing. Hind wings without spots. Fringes fuscous gray, dark brown at the ends of the veins.

Under side of fore wings about the same color as the upper, but shaded with darker brown at the base, and sprinkled with pale blue scales on the outer part. The hind wings colored as above, but with two irregular bands across them, limited by wavy black lines, and sprinkled on the outer part with pale blue lines. Body above and below dark brown.

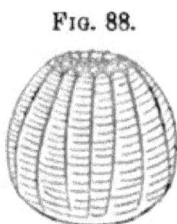

FIG. 88.

E. Pylades, egg, ×28.

Fig. 88 represents the egg of this species magnified twenty-eight diameters. The larva feeds on clover, the perfect insect being found in June, or earlier in the Southern States.

New England to Florida, Dakota, Colorado, California.

192. EUDAMUS BATHYLLUS, Sm.—Abb.

Expanse of wings from 1.4 to 1.5 inches.

Upper surface dark brown, about the same color as *E. Pylades*, with a slight grayish tinge. This is almost an exact copy of *E. Pylades* with the spots enlarged. The spots in the middle of the costa of the fore wings are connected with one in the cell that extends from the subcostal to the median vein, hour-glass-shaped, sometimes separated in the middle into two triangular spots. These are in line with two of the three spots below, forming a

y

triangle, the one in the second median interspace and the one below the lower branch of the median, the spot in the first median interspace being outside the line. The middle one of these last three spots is as large as the one in the cell, but is not so much constricted in the middle; the upper is next in size, and the lower one is small. The anteapical series consists of three quadrate, white, hyaline spots, with a white spot next the costa. Hind wings without spots, but with some brown clouding in the middle. Fringes pale gray, brown at the base, cut with brown at the ends of the veins on the fore wings.

Under side as in *E. Pylades*, except that there is more gray on the hind margin of the fore wings, and the white spots are larger than in that species, being a little larger than on the upper side. Body and head above concolorous with the wings; below grayish, the head and palpi whitish.

West Virginia to Florida, Illinois, Kansas, New Mexico.

193. EUDAMUS LYCIDAS, Sm.—Abb.

Expanse of wings from 1.9 to 2 inches.

Upper surface dark brown, slightly yellow-tinted, with a purple reflection along the costa. The fore wings are crossed from near the middle of the costa in a direction towards the posterior angle by a yellow band of spots, consisting of the same spots as are found on *E. Bathyllus*. The spot in the cell and the one in the second median interspace are greatly enlarged, so as to be nearly or quite quadrate; the others also are enlarged, but not to the same extent. Beyond this band there is the usual anteapical row of four spots, the fourth a little out of line, and, not in line with the others, a small triangular spot above the base of the first median venule. Hind wings without spots. Fringes of fore wings dark brownish gray, cut with black at the ends of the veins; those of the hind wings with the pale part paler; those of the inner margin black; a black line edging both wings, and shading in a little on the hind wings.

Under side of fore wings brown, blackish inside the mesial band, grayish along the hind margin, the apex shaded with black and having a few whitish scales. The spots of the mesial band are more confluent than above, the others less distinct.

Hind wings with the base dark, slightly grayish; through the middle an irregular broad black band, not reaching either margin, and having a large patch of brown in it at the end of the cell, the black sprinkled with gray scales. Outside this to the margin of the wing it is pure white through the middle half; the apical portion, the anal portion, and along the inner margin white more or less tinged with brown; the whole crossed by abbreviated brown streaks.

Body black, the under side of head and palpi slightly sprinkled with gray.

Massachusetts to the Gulf of Mexico, Mississippi Valley.

194. Eudamus Cellus, Bd.—Lec.

Expanse of wings from 1.9 to 2 inches.

Upper surface dark blackish brown; the fore wings crossed by a broad continuous yellow band, beginning near the middle of the costa and ending in a point near the posterior angle, bending inward a little here towards

the hind margin. From the costa to the lower branch
of the median the band is of nearly uniform width, but
the rest of the distance it tapers a little. The inner edge
is nearly straight, but the outer sends out an angle just
below the end of the cell. Three-fourths of the distance
from the base, extending from the costa back, is a short
anteapical line composed of three quadrate coalescing
yellow spots, and a small dot next the outer lower corner
of the third spot. Hind wings without marks, except a
little yellowish at the apex. Fringes black or blackish,
with a few gray or whitish spots between the veins.

Under side of fore wings the same as the upper, except that the hind margin is gray, the apical half of the
outer margin reddish brown, and the costal portion of
the mesial band and the anteapical line paler yellow.
Hind wings dark purplish brown, with three irregular,
somewhat poorly defined, darker brown bands, and a
very slight sprinkling of buff scales. Along the outer
margin is a series of deep brown lunules surrounded by
scattering pale blue scales.

Body black, under side of palpi and head pale ochraceous.

West Virginia to the Gulf of Mexico, Texas, Arizona.

195. EUDAMUS ZESTOS, Hüb.

Expanse of wings from 2.2 to 2.4 inches.

Upper surface dark brown, with a slight bronze reflection, marked almost like the upper surface of *E. Tityrus*.
This consists of a yellow band from near the middle of
the costa back towards the posterior angle, ending in a
blunt point about the middle of the medio-submedian
interspace, the point below the lower median fork being

about twice as large as in *E. Tityrus.* The inner edge of the band above the median fork is nearly straight, being notched a little at the subcostal vein, but is a little more oblique than in *E. Tityrus.* The outer edge is more irregular, the spot in the cell and the one in the lower median interspace being concave, and there are notches at the median and subcostal veins. The spot outside the band in the first median interspace is nearly square. Beyond the band is the usual anteapical line of three spots, extending obliquely outward in a curve.

Under side of about the same general color as the upper, the fore wings with the markings of the upper surface. The hind wings have a faintly-indicated median band of a slightly paler color than the rest of the wing, not reaching either margin, and two faint spots between this and the anal angle, and more or less scattering tawny scales. Both wings below have a strong purplish reflection, the outer margin slightly paler. Fringes concolorous with the wings, the hind wings paler, but without brown at the ends of the veins.

Body above purple-brown; thorax covered with tawny hairs, beneath more or less shaded with fulvous; palpi fulvous. Club of antennæ brown above, fulvous below and at the sides.

This strongly resembles *E. Tityrus* above, but the yellow spots are less confluent and more opaque, the general color is deeper, and the fringes are not divided by brown at the ends of the veins. The under side lacks the conspicuous silver band on the under side of the hind wings. This was described by Mr. C. E. Worthington as *E. Oberon.*

Florida ; Sanford, Marco Island.

196. EUDAMUS TITYRUS, Fab.

Expanse of wings from 1.8 to 2.1 inches.

Upper surface dark brown. The fore wings are crossed by an oblique yellow band of four large spots from the middle of the costa to near the posterior angle, where it ends in a rounded point, the lower part of the spot below the lower·median fork being but little narrower than the upper part which rests against the vein. The inner edge of this band is nearly straight, a little

FIG. 89.

Eudamus Tityrus, the left hand showing the under side of wings.

convex, the band narrowed a little as it approaches the costa. The outer edge is regularly dentate, the spot in the cell and the one below being straight on the outside. Beyond the band there is a narrow spot in the first median interspace extending from vein to vein. The ante-apical line near the costa is obliquely curved outward and composed of three spots. Fringes gray, cut with brown at the ends of the veins.

Under side brown, about the same shade as above, the outer and costal edges tinged with purplish gray. The

fore wings have the yellow spots of the upper side repeated, the spots more confluent. The hind wings have
a conspicuous silvery white band in the middle, nearly
reaching the costa, but not so near the inner margin.
This band is narrow in the anterior portion, but broadly
expanded in the middle, and rounded posteriorly, so as
to be somewhat flask-shaped. Palpi brown, slightly
yellowish beneath.

The larva of this species is to be found on the common
locust, rose acacia, Wistaria, and in the South on a species
of wild bean. When young, it cuts into the edge of a
leaf, and, drawing the flap over and fastening it with
silk, makes for itself a retreat, within which it stays.
As it increases in size, a larger section is cut in the leaf;
and, when this will no longer serve the purpose of protection, two or more leaves are fastened together. The
larvæ feed mostly at night, keeping themselves concealed
within their retreats during the daytime.

The egg is nearly globular, flattened at the base, with
fifteen ridges from base to apex; diameter, .04 of an
inch. Color white, with a bright red spot at the apex,
and a ring of the same color a little above the middle.
The duration of this period is about four days.

The young larva is .1 of an inch long, orange; the
head short, a little oblique, black; joint 2 dark brown;
a few hairs scattered over the body. Before the close of
this period the body shows a profusion of fine elevations.

In nine days from hatching the larva moults the first
time, when it is .2 of an inch long, with the colors the
same as during the preceding period, except that the
second segment is pale reddish brown, with a central
transverse dark brown stripe.

After the second moult, which occurs eleven days later, it is .55 of an inch long, the ground color yellowish black, with about six transverse yellow lines to each joint; the interspaces being dotted with yellow, giving the body a yellowish appearance. The last two segments are a little orange-tinted. Head cordate, this and the second segment brownish black, the latter shining, the head with an orange spot on each side above the ocelli.

In seven days more the larva moults again, when it is .75 of an inch long, with the color of the body unchanged; but the head is dark brown, and the second segment is black, with the sides and under parts red; the rest orange, the prolegs with a dark yellow base.

The larva moults the fourth time in four days more, when it is .9 of an inch long, but at the close of this stage, before pupating, it is 1.15 inches long. The head is broader than the middle of the body, a little oblique; and the second segment tapers anteriorly to a distinct neck. The body is a little flattened, tapering from the middle each way. The color remains about the same, the head assuming a little more of a wine color, and the top of the second segment and the jaws dark brown, the sides and feet on this joint about as before.

The time from the last moult to the change to a chrysalis varies with the season. One that moulted September 1 pupated October 4. This change takes place in the cluster of leaves it has woven together for a retreat during the larval period. Before pupating it lines the retreat with a thin coating of silk. The chrysalis is .76 of an inch long, the head-case

Fig. 90.

E. Tityrus, chrysalis.

blunt conical; the dorsum from near the head nearly straight, as shown in Fig. 90; the ventral side strongly ventricose, tapering abruptly from the end of the wing-cases to the tip of the abdomen, the cremaster a somewhat triangular piece, .06 of an inch long. Color reddish brown, finely mottled and spotted with dark brown, the end of the humerus blackish brown, with two smaller spots between them. Stigmata and eyes darker than the general color.

There are two or more broods of these butterflies in a season, the last brood hibernating in the pupa state, while the others emerge from the chrysalides in about two weeks.

United States generally.

197. Eudamus Proteus, Linn.

Expanse of wings from 1.9 to 2 inches.

Upper surface dark olive-brown; base of fore wings, basal half of hind wings, and upper part of body with light green hairs. The fore wings have an oblique transverse row of four whitish hyaline spots, extending from the costa near the middle to near the posterior angle. These spots are in the same position as the spots composing similar bands in *E. Zestos, Tityrus*, etc.; but they differ in being of about the same size and separated by the dark brown veins, the one in the cell constricted a little in the middle, the lower three near one another only at their corners. Beyond this band is a spot in the first median interspace, also constricted in the middle; and beyond the cell is an anteapical row of five spots, curved, the first two spots oblong, the third nearly quadrate, the fourth and fifth elongate in the direction

of the line, the **fourth** often divided in the middle into
two spots.

Hind wings without spots, the anal angle produced
into a tail .5 of an inch long, outer margin dentate.
Fringes of fore wings gray, cut with brown at the ends
of the veins; of hind wings white, except those of the
tail on the inner margin, which are black.

Under side of fore wings brown, the costa at base, the
area between the two bands, and the outer margin purplish glaucous. The spots are enlarged, and are more
confluent than above. The hind wings have the same
glaucous color all over their surface except the tail, the
anal two-thirds of the outer margin, and two bands
through the wings, which are olive-brown, the tail almost
black. The inner of these two bands reaches only to
the subcostal vein, and above that and a little to each
side are two black patches. The whole surface is
sprinkled over with a few whitish and yellowish scales.
Palpi pale gray, underneath almost white.

The mature larva is 1.5 inches long, fusiform, a fine
dark dorsal line, a bright yellow subdorsal band, which
is dilated on the twelfth segment, and a pale green line
along the base of the body. The dorsal space, between
the bands, is gray dotted with black and yellowish arranged in transverse lines. The sides are gray, with the
upper half dotted with black. Collar lustrous black;
anal plate yellow, greenish in the middle. Under side
pale green, legs black, prolegs yellow. Head large,
round, brown, pubescent, slightly depressed at top; a
yellow spot on each side of the mouth, narrowing upward, and fading into the light brown of the upper
part of the face.

Chrysalis.—Covered with a white powder. The larva feeds on leguminous plants,—*Phaseolus perennis* and *Clitoria Mariana.*

Southern States; occasionally in New York.

198. ERYCIDES BATABANO, Lef.

Expanse of wings from 2.3 to 2.6 inches.

Male.—Upper surface deep smoky brown, with a pronounced indigo-violet reflection. Fore wings without markings, other than a few scattering blue or green scales about the base of the wings. Fringes concolorous.

Hind wings with a row of brilliant blue or green elongated submarginal spots, more or less confluent, interrupted by the veins, and becoming obsolete towards the costal margin. Fringes with some white in the intervenular spaces.

Under side of fore wings paler, with a purplish cast, a few blue or green scales along the costa, about the base, and near the posterior angle. Hind wings much like the upper side, but deeper, the purplish reflection at the costal margin gradually changing to deep indigo as it approaches the abdominal folds; a few blue or green scales in the median space and along the inner margin. Submarginal spots as above, but brighter.

Female.—Upper side paler than in the male, gradually growing deeper over both wings until nearly black at the anal angle, both wings with a faint purplish reflection in certain lights. Markings on the hind wings like those of the male. Beneath differing from the male only in its paler color, and in having a purple reflection, which is more prominent than on the upper side.

Body deep brown above and below, some blue or green scales on the collar, front, and shoulders, and arranged in bands on the posterior segments of the abdomen. Palpi greenish white. Antennæ dark brown. Hind wings produced a little at the anal angle.

Florida.

199. ERYCIDES AMYNTAS, Fab.

Expanse of wings 2 inches.

Upper surface dark brown, almost black, with a strong purple reflection. A little beyond the middle of the cell of the fore wings is a white hyaline bar extending across the cell, emarginate externally. Below this, and a little farther out, in the second median interspace, is a similar but larger subquadrangular spot, and in the first median interspace is another, more oblong in shape. The first two of these three spots represent two of the four spots which form the median oblique band of *Eudamus Tityrus* and allied species. There is near the apex an oblique row of three anteapical small spots, subquadrate in form, the first the smallest. Hind wings without spots. Fringes fuscous, those of the fore wings darkest; the anal angle somewhat produced.

The under side has the spots of the upper surface of the fore wings repeated. The surface along the costa, a broad apical portion and the external margin of the fore wings, and all of the hind wings, except two bands of spots and the anal angle, rich purple, not very dark; all the rest of the surface is dark brown. The dark brown portion consists of two transverse bands, marking nearly the division of the wing into thirds, and a broad portion around the anal angle. Body black; palpi and under side of head gray.

Key West, Florida.

200. MEGATHYMUS YUCCÆ, Bd.—Lec.

Expanse of wings from 2.5 to 3 inches.

Male.—Upper surface deep umber-brown, the base of both wings tinged with yellow, the markings yellow.

FIG. 91.

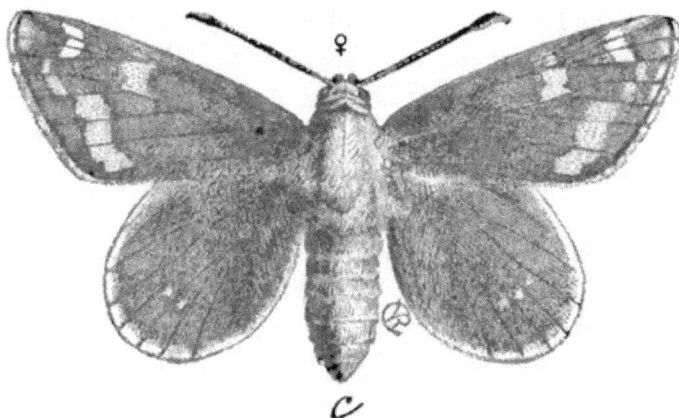

Megathymus Yuccæ, female (natural size).

The fore wings have a large spot in the outer end of the cell that is subquadrate; above this in the interspaces are three small spots in an oblique line, but little more than mere dots. Beyond these is a submarginal row of spots which begin in the usual line of anteapical spots about four-fifths the distance from the base to the apex, but the two spots opposite the cell are nearer the outer margin, and from these they gradually run nearer the margin, till the last one is close to the posterior angle, as shown in Fig. 91. The first of the four anteapical spots is a mere dot, the next three are a little longer than wide, the fifth and sixth are narrow but reaching from vein to vein, the seventh and eighth are in the median interspaces, and the ninth is above the submedian. The last three are somewhat irregularly convex on the inner side, but less so on the outer. Hind wings without spots, but with a yellow washing along the outer margin.

Under side deep brown, like the upper, but brighter, the outer portion of both wings pearly gray, with a patch of the same color on the middle of the costa of the hind wings, and a white crescent below it, and the same scales sprinkled along the inner margin and the faint streaks through the wings. The spots of the fore wings are repeated, but somewhat enlarged, the color paler yellow, whitish in the costal region.

Female.—The general color and markings of the fore wings as in the male, but much larger. The spot in the cell extends from vein to vein, and inward along the median a little way towards the base, the three spots above nearly as long as the spot in the cell is wide. Besides the much enlarged anteapical spots, which are paler yellow than the other spots, there are three oblique pale yellow spots on the costa before the apex. The lower three spots of the submarginal band are widened, so that the inner upper corner almost reaches the cell. Hind wings, besides the yellow scales along the margin, have a discal row of four small spots not reaching either margin. Under side as in the male, except that the spots are more prominent, and the spots near the costa of the fore wings are more washed with white. The female is larger than the male, the smaller of the expanses given applying to the male.

According to Professor C. V. Riley, the larva of this species lives inside the stem and root of the Yucca, or Spanish Bayonet, being the only one of our butterflies that lives in the larval state as a borer on the inside of the stems of plants, unless we include the allied species *M. Cofaqui.* There is a probability that when the life-history of that species is known it will be found to have similar habits. The eggs are deposited singly on the leaves, and, when hatched, the larva conceals itself in a web between some of the more tender terminal leaves. Generally it will be found at first near the tip of a leaf, where the sides naturally roll up and afford a safe retreat. It then gradually works to the base, feeding as it goes, and rolling and shrivelling the blade as it descends. Other blades are often joined, the insect living among the blades till it is about one-fourth grown, seldom entering the stem before that time.

The egg is subconical, the top flattened or depressed and with a slight central dimple, the base concave, smooth but not polished. Color pale green when first deposited, but inclining to buff-yellow or brown before hatching. The diameter is about .1 of an inch, the height about .06.

The young larva is about .2 of an inch long, of a dark brick-red color, head and top of the second segment pitchy black. The abdominal joints show two principal transverse folds. There are six rows of stiff black hairs arising from the body or from very small tubercles. Head larger than the second segment, rounded, but somewhat flat in front; cervical shield narrow and in one piece; both minutely punctured. No anal plate.

The full-grown larva averages 2.6 inches in length by

.4 of an inch in diameter.   Color dull translucent white.
Surface faintly aciculate, sparsely armed dorsally with
minute, evenly-distributed, short, rufous bristles, springing from the general surface, and not very noticeable
with the naked eye; covered more or less copiously with
a white, glistening, powdery secretion of a waxy nature.
Cylindrical; the abdominal joints with eight annulate

FIG. 92.

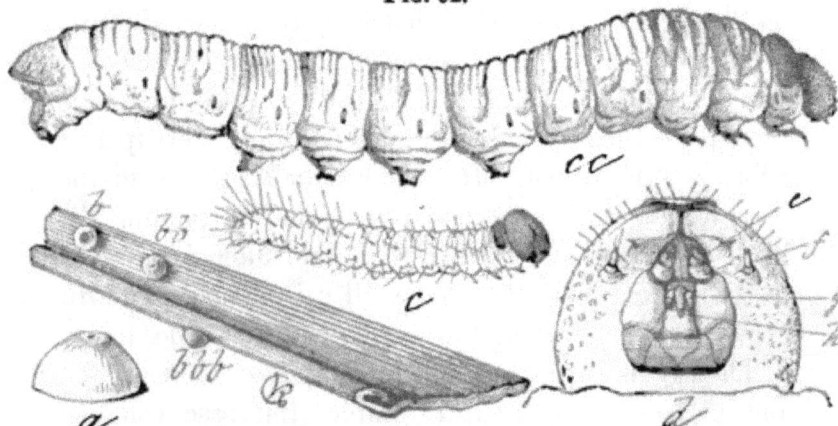

Megathymus Yuccæ: *a*, egg, side-view, enlarged; *b*, egg from which the larva has
hatched; *bb, bbb*, unhatched eggs, natural size; *c*, newly-hatched larva, enlarged;
cc, full-grown larva, natural size; *d*, under side of head of same, enlarged to show
the trophi.

or transverse wrinkles to each joint,—the first three occupying the anterior half, the third more prominent and
widening laterally, and the other five on the posterior
half of the joint, all best defined dorsally.   The thoracic
joints somewhat larger than the rest, more deeply and
irregularly wrinkled; the substigmatal region with longitudinal folds.   Head black, perpendicular, and aspe-
rous or deeply shagreened; epistoma and labrum brown,
small, and usually with a transverse median ridge, the

λ-shaped mark white, forking before the suture, the forks
having the shape of U; mandibles stout, subtriangular,
non-dentate; antennæ (Fig. 92, f) two-jointed, exclusive
of the bulbus, the terminal joint twice as long as the
basal; maxillæ and labium and mentum forming a subquadrate piece, bulging out prominently from beneath,
the parts seemingly soldered together and separated only
by deep sutures; the maxillary palpi (Fig. 92, e) consisting of two broad joints; the labium small, trapezoidal,
highly polished, with the spinneret (h) twice as long as
the palpi (g), which are small, recurved, and two-jointed,
exclusive of the bulbus; a few stout bristles on the
labrum, on the palpigerous piece of the maxilla, on the
mentum, on the base of the mandibles, and around the
ocelli, which are not easily distinguished from the more
globular of the shagreenations. Cervical shield more
glabrous than the head, and scarcely darker than the
body, except around the hind border. Thoracic legs
very short, but stout, with the horny parts deep brown,
and sparsely armed with bristles. Prolegs well developed, the hooks in a double row, and forming a distinct
purple-brown, transversely oval annulus, but slightly
broken at the narrow ends. Anal shield rounded behind, coriaceous rather than corneous, and with a slight
increase of bristly hairs, especially around the border.
Stigmata large, with a purple-brown, oval annulus.

Chrysalis.—Average length 1.5 inches; cylindrical,
broadest at the shoulders, recurving ventrally towards
the tip, and terminating in a broad, flattened, posteriorly
rounded, transverse, slightly decurving flap, the borders
thickened basally and extending ventrally. Eyes prominent, with a transverse carina; wing-cases reaching the

hind part of the fourth abdominal segment. Surface but slightly polished, and faintly corrugate; a few extremely minute, bristle-like spines distributed over the abdominal joints dorsally, and the two or three terminal joints with stiff rufous hairs, increasing posteriorly. Color black-brown anteriorly, paler on the abdomen, and more or less densely covered with a white powdery secretion like that on the full-grown larva.

FIG. 93.

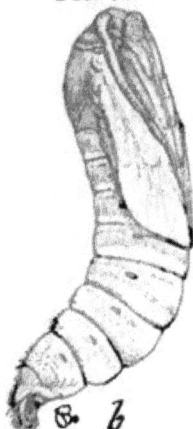

Megathymus Yuccæ, pupa.

There is but one brood in a season, the butterflies appearing in April and May. The eggs hatch in about ten days, but the larva does not pupate till late in the following winter or early in the spring. The burrow often extends two feet or more below the surface of the ground. Before pupating, the larva makes a place of exit for the butterfly, lightly closing this cavity near the top. It then makes a cell sufficiently farther down to give it room enough to pupate, and in this it undergoes its transformations.

Southern States, New Mexico, Arizona.

201. MEGATHYMUS COFAQUI, Strecker.

Expanse of wings from 2.25 to 2.6 inches.

This differs from *M. Yuccæ* in having more yellow at the base of the wings, the female having the yellow spots in and above the cell of the fore wings connected in a continuous row with those below instead of the spots beyond the cell, and the anteapical spots making up with these the row across the wing. This is due to the greater expansion inwardly of the three lower spots, so as to bring them under the cell.   Below the median vein there is a yellowish spot about one-third the distance from the base to the margin.   The anteapical spots and the two beyond the cell are like those in *M. Yucca.*   Hind wings with a yellow spot on the costa and a more distinct yellow border, besides the discal row of yellow spots.

The male has on the outer half of the fore wings an irregular row of five pale yellow spots of various sizes and shapes, the lower three pointed internally.   The hind wings have a rather narrow, even, straw-colored border.   Fringes dirty white.

The under side of the fore wings of both sexes is like the upper, the hind wings with the addition of some gray shading on the costa and several subterminal white spots, otherwise much as in the other species.

The whole of the upper surface of the hind wings of the male, except the space occupied by the outer border, is thickly covered with long, fine, black hairs, which stand nearly at right angles to the surface, to the height of nearly a quarter of an inch.   The basal third of the under side of the fore wings is furnished with a similar coat of hairy appendages.

Florida, Colorado.

GLOSSARY.

Abdomen, the posterior part of the body.

Aciculate, needle-shaped, more slender than subulate.

Alba, or *Albus*, white.

Anal, pertaining to the posterior part of the body.

Anal angle, the inner or posterior angle of the hind wings, next to the body.

Annulate, furnished with colored rings.

Annulus, a ring.

Anteapical, before the apex; on the front or costal portion of the wing, near the apex.

Antennæ, two articulated organs of sensation, situated on the head.

Anterior margin, the front margin of a wing; the costa.

Apex, that part of a wing which is farthest from the body; the angle between the costa and the outer margin.

Approximate, near to; near together.

Aureus, golden yellow.

Azure, sky-blue:—nearly the same as *cæruleus*.

Basal, relating to the base.

Base, as applied to a wing, that part which is joined to the body.

Bifid, cleft.

Bipupillate, applied to an ocellate spot having two pupils or dots within it of a different color from the rest of the spot.

Blind, applied to a round spot destitute of a central spot or pupil.

Body, the three parts of an insect,—head, thorax, and abdomen.

Bronze, the color of old brass.

Cæruleus, color of the sky; sky-blue.

Cæsius, pale blue, approaching gray.

Canus, hoary, with more white than gray.

Capillary, hair-like; long and slender, like a hair.

Carina, a keel.

Carinate, keeled; having a longitudinal prominence like the keel of a boat.

Carneous, flesh-colored.

Caterpillar, the larva.

Cauda, the tail.

Caudate, tailed:—generally applied to the posterior wings of Lepidoptera to indicate tail-like projections.

Cell, usually in Lepidoptera the space between the subcostal and median veins; the discal cell.

Chitine, the hard part on the outside of insects.

Chrysalis, the third stage of **the** insect, counting the egg one.

Ciliæ, fringes.

Ciliate, fringed.

Cinereous, ash color; gray tinged a little with blackish; the color of wood ashes.

Cingula, a colored band.

Clavate, club-shaped.

Coalesce, to grow together or unite.

Collar, scales back of the head, forming more or less of a ring; the neck.

Concolorous, of the same color, as the body agreeing in color with the wings.

Confluent, running into each other.

Connivent, converging or approaching.

Contiguous, touching; placed so near as to touch.

Convergent, approaching each other towards the tip.

Cordate, heart-shaped.

Corrugated, **wrinkled.**

Costa, the thickened anterior part of the wing from the base to the apex.

Coxa, the jointed base of the leg.

Cremaster, the anal hooks of the chrysalis, which fasten into silk to hold the chrysalis in place.

Crenate, scalloped.

Cuneiform, wedge-shaped.

Cupreous, coppery; the color of copper.

Cyaneus, dark blue, like Prussian blue.

Dentate, toothed.

Diaphanous, semi-transparent; clear.

Dichotomous, forked; dividing by pairs.

Diffuse, spreading.

Dimorphous, two-formed :—applied to a species existing in two forms having different colors or markings.

Disk, the surface within the margin,—usually between the end of the cell and the outer margin.

Dorsal, of the back.

Dorsum, the back or upper surface.

Echinate, set with prickles.

Edematous, dull translucent white.

Edge, the margin.

Egg, the first stage of an insect.

Elliptical, in the form of an ellipse.

Emarginate, notched.

Entire, the margin smooth, or without teeth.

Epupillate, applied to an ocellate spot included in a colored ring, but destitute of a pupil or central dot.

Erect, upright.

Eyes, the organs of sight, composed of numerous hexagonal facets.

Face, the anterior or front part of the head.

Fascia, a transverse band or broad line.

Fasciated, banded.

Feet, the organs of motion.

Femur, the thigh or third part of the leg.

Ferruginous, of the color of iron-rust.

Filiform, thread-shaped.

Flexuous, zigzag without acute angles.

Fuliginous, sooty ; soot-colored.

Fulvous, orange-yellow.

Fuscous, dark brown with a slight mixture of gray.

Fusiform, spindle-shaped; gradually tapering towards each end from near the middle.

Geminate, situated in pairs.

Genus, an assemblage of species which correspond in particular characters.

Glabrous, smooth.

Glaucous, gray bluish green.

Globular, like a round ball.

Glutinous, slimy, viscid.

Granulated, covered with small grains.

Gregarious, living in society, or many feeding together.

Griseous, light gray

Habitation, or *Habitat,* a **situation** or locality **frequented by insects.**

Head, the anterior part **of the** body.

Hibernaculum, a case of web and leaves in which larvæ or **pupæ** hibernate; or a cocoon of silk.

Hibernate, to **pass** through or survive the winter.

Hind margin, that part of the fore wings which is included between the base and the posterior angle.

Hirsute, rough with strong hairs.

Hispid, bristly; rough with stiff, short, sparse hairs.

Hoary, covered with a fine white silvery substance or pubescence.

Humerus, the anterior **base of** the wing.

Hyaline, **transparent; vitreous.**

Imago, the **perfect or adult insect.**

Imbricated, **tiled; placed one over another, like shingles on the roof of a house.**

Immaculate, **without spots.**

Incanous, hoary.

Inconspicuous, not readily discernible.

Inner margin, or *Interior margin,* **that margin of the hind wings which extends from the base to the anal angle; by some authors used to denote the posterior or hind margin of the** fore wings.

Iris, of an ocellate wing-spot, is a circle that surrounds the principal spot.

Irrorate, sprinkled.

Joints, or *Articulations,* the divisions of the body or segments of the larva; **the divisions** of the **pupa,** more particularly the abdomen; the **divisions of the** antennæ.

Keel, the carina.

Labial palpi, articulated filaments, one on each side of the labium.

Labium, the lower lip.

Labrum, the upper lip.

Lanceolate, lance- or spear-shaped.

Larva, the second stage of an insect, counting the egg the first.

Larvarium, a retreat of silk and leaves, or of silk, in which some larvæ stay when not feeding.

Lateral, situated on the side.

Lenticular, lens-shaped.

Lepidoptera, an order of insects having four wings covered with minute imbricated scales; butterflies and moths.

Lethargic, becoming torpid or inactive.

Ligula, tongue.

Lilacinous, lilac color.

Linear, narrow and of nearly uniform width.

Lineated, streaked, or marked with lines.

Livid, dark gray, verging towards violet.

Longitudinal, the direction of the longest diameter.

Lunate, crescent-shaped; formed like a new moon.

Lurid, of a dirty brown color.

Luteus, unmixed yellow.

Macula, a spot larger than a puncture, of some other than the general color.

Maculate, or *Maculated,* spotted.

Mandibles, the upper jaws.

Margin, the edge of a wing, or along the edge.

Maxillæ, the lower jaws, placed between the upper jaws and the lower lip.

Maxillary palpi, filaments attached to the maxillæ.

Mesial, middle, as a band or stripe across the middle portion of the wing.

Mesonotum, the covering of the middle of the dorsal portion of the thorax.

Mesothorax, that division of the thorax to which the middle pair of legs are attached.

Metamorphoses, transformations.

Metathorax, that division of the thorax to which the hind pair of legs are attached.

Micropyle, the apex of the egg of an insect.

Moult, or *Molt,* shedding or casting off the larva skin.

Nebulous, clouded.

Nervures, divisions of the nerves or veins of a wing.

Niger, black a little tinged with gray.

Obconic, inversely conic.

Obcordate, inversely heart-shaped.

Oblong, the transverse diameter much shorter than the longitudinal.

Obovate, inversely egg-shaped.

Obsolete, indistinct.

Occiput, the hinder part of the head.

Ocelli, eye-like spots on the wings of Lepidoptera; simple eyes of insects.

Ochreous, yellow **with a slight** tinge of **brown.**

Orbicular, round.

Order, the subdivision **of a** class.

Osmateria, scent-organs **of the** larvæ of the genus Papilio.

Oval, broadly elliptical.

Ovate, egg-shaped.

Overlaid, heavily sprinkled with scales of a different color from the ground color; clouded; overcast.

Palpi, in butterflies, the three jointed organs beneath the head between which the tongue is coiled like a watch-spring.

Piceous, pitchy; the color of **pitch.**

Pile, very minute, **short hairs.**

Pilous, having **long, sparse hairs.**

Polymorphous, applied **to a species existing in several different forms as to color, markings, or size.**

Porrect, **straight out.**

Posterior angle, **the angle formed by the outer margin** and the posterior or hind **margin of** the fore wing.

Posterior margin, **that portion of the fore** wings which is opposite **the costa.**

Proboscis, **the** tongue, or sucking organ.

Prolegs, **the fleshy legs of** caterpillars.

Pronotum, the anterior part of the covering **of the** thorax; the **covering** of the thorax.

Prothorax, the first division of the thorax, to which the first pair of legs are attached.

Pruinous, **hoary;** covered **with a whitish powder.**

Pubescent, **coated with** fine **hair or down.**

Punctured, **marked with small impressed dots.**

Pupa, **the third stage of an** insect, counting **the egg the first; the chrysalis.**

Pupate, **to** assume **the pupa form.**

Pupil, **of an** ocellus, the central point.

Quadrangular, **having** four angles.

Quadrate, square, or nearly square.

Remote, separate; not near together.

Reniform, kidney-shaped.

Reticulate, resembling net-work.

Retractile, capable of being exserted or drawn in at pleasure.

Retuse, ending in an obtuse sinus.

Ribs, ridges on eggs from the base to the apex.

Roseus, rose color.

Rosy, rose color.

Rufous, reddish.

Rugous, or *Rugose*, wrinkled.

Sagittate, arrow-shaped.

Sanguineous, of the color of arterial blood.

Scabrous, rough, with projecting points.

Scales, the dust or imbricated pieces covering the wings.

Segment, a ring or division of the body.

Sericeous, silky.

Serrate, saw-toothed.

Sessile, connected with the part to which it is attached without the intervention of a peduncle or stalk.

Seta, a bristle.

Sexes, the two divisions of animals : in insects distinguished by ♂ for male and ♀ for female.

Sinuate, indented.

Sinus, an indentation or excavation.

Sparse, scattered.

Species, an assemblage of individuals possessed of permanent characters of size, color, and ornamentation, by which they may be distinguished from other forms, and which breed true to their type.

Spinous, armed with spines.

Spiracles, breathing-holes on the side of the body ; the stigmata.

Sprinkled, marked with thinly-scattered scales of another color than the ground color.

Stigmata (singular, *Stigma*), the breathing-holes on the sides of the body ; also, sometimes, spots on a wing.

Striæ, lines ; transverse elevated lines on eggs.

Striate, marked with lines.

Submarginal, applied to a space or line within the margin.

Subocellate, applied to an ocellus without a pupil.

Suffused, blurred with a color other than the usual one.

Sulphureous, bright yellow ; the color of sulphur.

Tail, the terminal segment of the abdomen ; an appendage at the posterior part of the hind wings.

Tarsi, the feet.

Tawny, dull yellowish brown.

Terminal, at the extremity.

Testaceous, tile or brick color.

Thorax, that part of the body **which** is back of the head.

Tibia, that part of the leg **which** is next to the foot.

Tongue, **the** sucking-tube of Lepidoptera.

Torpidity, a lethargic state of hibernation.

Transverse, crosswise.

Trochanter, an appendage at the base of the thigh.

Trophi, the mouth parts.

Truncate, cut square off.

Tubercle, a small swelling **or prominence.**

Variety, a form of one **or more examples of a** species differing from the usual form, but not breeding true to type.

Veins and **Venules,** the framework of the wings.

Venter, the lower part of the body.

Ventricose, distended.

Villi, **soft hairs.**

Violaceous, violet color.

Vitellinus, yellow with a slight tinge of red.

Vitta, a longitudinal colored line.

Washed, covered with **scales of a color different from the ground color,** but not quite obscuring the latter.

INDEX.

THE END.

VALUABLE WORKS OF REFERENCE.

Lippincott's Pronouncing Biographical Dictionary.

Containing Complete and Concise Biographical Sketches of the Eminent Persons of all Ages and Countries. *New Edition, Thoroughly Revised and Enlarged.* By J. THOMAS, M.D., LL.D. 1 vol. Imperial 8vo. Sheep. $12.00.

Allibone's Critical Dictionary of Authors.

A Dictionary of English Literature and British and American Authors, Living and Deceased. By S. AUSTIN ALLIBONE, LL.D. 3 vols. Imperial 8vo. Extra cloth. $22.50. Sheep. $25.50.

Lippincott's Pronouncing Gazetteer of the World.

A Complete Geographical Dictionary. New Edition of 1880. Thoroughly Revised. *Containing Supplementary Tables,* with the most recent Census Returns. Royal 8vo. Sheep. $12.00.

Allibone's Dictionary of Prose Quotations.

By S. AUSTIN ALLIBONE, LL.D. With Indexes. 8vo. Extra cloth. $3.00. Cloth, gilt. $3.50. Half calf, gilt. $5.00.

Allibone's Dictionary of Poetical Quotations.

By S. AUSTIN ALLIBONE, LL.D. With Indexes. 8vo. Extra cloth. $3.00. Cloth, gilt. $3.50. Half calf, gilt. $5.00.

Chambers's Encyclopædia.

Revised Popular Edition.

A Dictionary of Useful Knowledge. Profusely Illustrated with Maps, Plates, and Wood-cuts. 10 vols. Royal 8vo. Cloth. $25.00.

Chambers's Book of Days.

A Miscellany of Popular Antiquities Connected with the Calendar. Profusely Illustrated. 2 vols. 8vo. Extra cloth. $8.00.

Dictionary of Quotations.

From the Greek, Latin, and Modern Languages. With an Index. Crown 8vo. Extra cloth. $2.00.

Furness's Concordance to Shakespeare's Poems.

An Index to Every Word therein contained, with the Complete Poems of Shakespeare. 8vo. Extra cloth. $4.00.

Lempriere's Classical Dictionary.

Containing all the Principal Names and Terms Relating to Antiquity and the Ancients, with a Chronological Table. 8vo. Sheep. $2.60.† *Abridged.* 12mo. Cloth. $1.08.†

☞ The above Works are also bound in a variety of handsome extra styles.

Allibone's Dictionary of English Literature,

And British and American Authors, Living and Deceased. From the Earliest Times to the Middle of the Nineteenth Century. Containing over 46,000 Articles (Authors), with 40 Indexes of Subjects. 3 vols. Royal 8vo. Extra cloth, $22.50. Sheep, marbled edges, $25.50. Half calf, gilt, $30.00. Half morocco, Roxburgh, gilt top, $31.50. Half Russia, $33.00.

The Reader's Handbook

Of Facts, Characters, Plots, and References. By E. Cobham Brewer, LL.D. 12mo. Half morocco, $3.50. Sheep, $4.00.

Reader's Reference Library.

THE READER'S HANDBOOK.
WORDS, FACTS, AND PHRASES.
ANCIENT AND MODERN FAMILIAR QUOTATIONS.
WORCESTER'S COMPREHENSIVE DICTIONARY.
ROGET'S THESAURUS.

5 vols. Half morocco, in cloth box, $12.50.

Baily's Trees, Plants, and Flowers.

Where and How they Grow. With 73 Engravings. 16mo. Extra cloth, 75 cents.

Baily's Our Own Birds of the United States.

A Familiar Natural History. Edited by Ed. D. Cope. To which is added Baily's Trees, Plants, and Flowers. With numerous Illustrations. 12mo. Toned paper. Cloth extra, $1.25.

Cowan's Curious Facts in the History of Insects, including Spiders and Scorpions; together with their Uses in Medicine, Art, and as Food; and a Summary of their Remarkable Injuries and Appearances. 12mo. Cloth, $2.25.

Cozzens's Sparrowgrass Papers;

Or, Living in the Country. New Edition. 12mo. Cloth, $1.75.

Buckland's Log-Book of a Fisherman and Zoologist. Illustrated. Crown 8vo. Extra cloth, $3.00.

Brisbin's The Beef Bonanza;

Or, How to Get Rich on the Plains. Being a Description of Cattle-Growing, Sheep-Farming, Horse-Raising, and Dairying in the West. Illustrated. 12mo. Extra cloth, $1.25.

Howard's Fifty Years in a Maryland Kitchen.

A Complete Cook-Book. New Edition. 12mo. Fine cloth, $1.50.

Harland's Farming with Green Manures;

Or, Plumgrove Farm. Second Edition, Revised and Enlarged. 12mo. $1.00.

Up de Graff's Camping Out;

Or, Bodines. A Complete Practical Guide to "Camping Out." Profusely Illustrated. 12mo. Extra cloth, $1.25.

Prantl's An Elementary Text-Book of Botany.

Translated from the German. The translation revised by S. H. Vines, M.A., D.Sc., F.L.S. With 275 Illustrations. 8vo. Extra cloth, $2.25.

Nystrom's Pocket-Book of Mechanics and

Engineering. Eleventh Edition. Illustrated. 16mo. Pocketbook form. $3.50.

Sloan's Homestead Architecture.

Containing 40 Designs for Villas, Cottages, and Farm-Houses. With Essays on Style, Construction, Landscape Gardening, Furniture, etc. Illustrated with upwards of 200 Engravings. New Edition. 355 pages. 8vo. Cloth, $3.50.

Hobbs's Architectural Designs for Country

and Suburban Residences. With upwards of 100 Engravings. New, Revised Edition. 8vo. Cloth extra, $3.00.

Mayhew's Illustrated Horse Management.

Containing Descriptive Remarks upon Anatomy, Medicine, Shoeing, Teeth, Food, Vices, and Stables. Likewise, a Plain Account of the Situation, Nature, and Value of the Various Points, together with Comments on Grooms, Dealers, Breeders, Breakers, Trainers, Carriages, and Harness. Embellished with more than 400 Engravings from original designs made expressly for this work. 8vo. Cloth, $3.00.

Mayhew's The Illustrated Horse Doctor.

An Accurate and Detailed Account of the Various Diseases to which the Equine Race is Subject; together with the Latest Mode of Treatment, and all the Requisite Prescriptions Written in Plain English. Illustrated with more than 400 Engravings. 8vo. Cloth, $3.00.

Dwyer's Horse Book.

On Seats and Saddles, Bits and Bitting. By Francis Dwyer.
Illustrated. 12mo. Extra cloth, $1.25.

Youatt on the Horse.

Its History, Treatment, and Diseases. With numerous Illustrations. 8vo. Cloth extra, $2.00.

Youatt on the Dog.

Edited, with Additions, by E. J. Lewis. With 22 Illustrations. 1 vol. 8vo. Extra cloth, $2.50.

Wharton's Treatment of the Horse.

A Hand-Book on the Treatment of the Horse in the Stable
and on the Road, or Hints to Horse Owners. With many
Illustrations. 12mo. Cloth extra, $1.25.

Lewis's The American Sportsman.

Containing Hints to Sportsmen, Notes on Shooting, and the
Habits of the Game Birds and Wild Fowl of America.
New Edition, Revised and Enlarged. With New Chapters
by Arnold Burges. Illustrated. 8vo. Extra cloth, $2.50.

Oswald's Zoological Sketches.

A Contribution to the Outdoor Study of Natural History.
By the author of "Summerland Sketches." With 36 Illustrations by Hermann Faber. Cloth gilt, $2.50.

Rohrer's Practical Calculator.

A Pocket Manual of Plain Rules and Calculations for Business Operations. 18mo. Half bound, 50 cents.

Hanna's Ready Reckoner.

Complete Ready Reckoner, and **Log, Table, and** Form Book.
(U. S. Standard.) *New Edition, Revised and Enlarged.*
16mo. Half bound, **60 cents.** Cloth, 75 cents.

Ready Reckoner and Mercantile Companion.

Pocket Edition. 32mo. Limp cloth, 25 cents.

**Any of the books in this list will be forwarded by mail or express
transportation free, on receipt of the price by the Publishers,**

J. B. Lippincott Co.,

715 and 717 Market St., Philadelphia

𝕸𝖎𝖘𝖈𝖊𝖑𝖑𝖆𝖓𝖊𝖔𝖚𝖘.

Rand.

Elements of Medical Chemistry. 12mo, cloth, $2.00.

Ad. Wurtz.

Elements of Modern Chemistry. Illustrated. *New and Enlarged Edition.* 12mo, cloth, $2.50; sheep, $3.00.

Contemporary Science Library.

Vol. I. The Science of Language. By Abel Hovelacque. 12mo, cloth, $1.50.

Biology. By Dr. Charles Letourneau. Illustrated. Crown 8vo, cloth, $1.50.

Anthropology. By Dr. Paul Topinard. Crown 8vo, $1.50.

Æsthetics. By Eugene Veron. 12mo, cloth, $1.50.

Philosophy. By A. Lefèvre. 12mo, cloth, $1.50.

J. H. Hobbs.

Country and Suburban Residences. With upwards of 100 Engravings. 8vo, cloth, $3.00.

Samuel Sloan.

City and Suburban Architecture. Imperial 4to, cloth, $10.00.

Constructive Architecture. Quarto. Cloth, $7.50.

Homestead Architecture. 8vo, cloth, $3.50.

The Model Architect. 2 vols. Imperial 4to, half morocco, $18.00.

Addison.

Complete Works. 6 vols. 12mo, cloth, $7.50.

Thomas à Kempis.

On the Imitation of Christ. Beautifully Illustrated and elegantly printed on superfine paper. 12mo, cloth, gilt top, $5.00. 16mo, cloth, $1.00.

Critical **Commentary.**

A Commentary, Critical, Experimental, and Practical, on the Old and New Testaments. By Drs. Brown, Jamieson, and Fausset. With 16 Maps and Plans. In 6 vols. Royal 8vo, cloth, $15.00.*

Foreign Classics for English Readers.

Edited by Mrs. Oliphant. 16mo. Per vol., cloth, $1.00.

Dante. By Mrs. Oliphant.

Voltaire. By Col. E. B. Hamley, C.B.

Pascal. By Rev. Principal Tulloch.

Goethe. By A. Hayward, Esq., Q.C.

Petrarch. By H. Reeve, Esq., C.B.

Montaigne. By Rev. W. L. Collins.

Molière. By F. Tarver.

Tasso. By E. J. Hasell.

Rabelais. By W. Besant.

Cervantes. By Mrs. Oliphant.

Calderon. By E. J. Hasell.

St.-Simon. By C. W. Collins.

Corneille and Racine. By H. M. Trollope.

Madame de Sévigné.

La Fontaine, etc. By Rev. W. L. Collins.

Schiller. By J. Sime.

Rousseau. By H. Graham.

Other Volumes in Preparation.

Goethe.

Complete Works. 11 vols. 12mo, cloth, $11.00.

Hahn.

Hebrew Bible. 8vo, half morocco, $3.15.*

C. Harlan, M.D.

Farming with Green Manures; or, Penngrove Farm. *Second Edition, Revised and Enlarged.* 12mo, cloth, $1.00.

Mrs. B. C. Howard.

Fifty Years in a Maryland Kitchen. 12mo, cloth, $1.50.

www.ingramcontent.com/pod-product-compliance
Lightning Source LLC
Chambersburg PA
CBHW021350210326
41599CB00011B/822